When Logging Was Logging

100 Years of Big Timber in Southwest Washington

DONORS AND SUPPORTERS

Anita Raistakka
Annette Boslar
Anonymous Donors (2)
Brian and Linda Elliott
Burton Arthur Appelo
Carlton Edward Appelo
Cecilia R. Smith
Cowlitz River Rigging
Cummins NW
Dan and Paula Cothren
David and Christy Nicandri
Delores Nicholson Carlson
Dennis and Carol Larson
Dianna Jordan
Don Raistakka
F. L. Delay
Frederick and Sadie Lehr
Gary and Susan Burkhalter
George and Donna Klint
Gerald Rasmussen
Glenn and Sharon Raul
Grays River Grange #124
Howard and Arlene Nelson
In memory of C. A. and Agnes Appelo
In memory of Eric Nicholson
In memory of Grant Erickson
In memory of Joe "Hooly" Souvenir
In memory of Jack and Mark Linquist
In memory of RV "Duffy" Larson
In memory of Werner Nikkila
Jerry DeBriae Logging Company
Jerry R. Wirkkala
John and Jane Scharbach
Karen Elliott Bertroch
Keith and Marlys Metcalf
McPaul Cat and Crane
Merle and Bettie Sorenson
Miller Family Foundation
Nancy Bell Anderson
Norman and Myrtle Anderson
Peter A. Ringen
Peter Brix
Philip and Mindy Penttila
Randolph and Wanda J. Kelly
Rich and Randee Erickson
Robert W. Anderson
Rodney and Marilyn Madden
Sandra Lopez
Scott and Colleen Smith
Senator Sid and Mrs. Betty Snyder
Shirley Cooper
Sorenson Investment Group
Sorenson Transport
Theresa Appelo Bakken
Todd and Karen Kaino
Watkins Tractor Company
Western Wahkiakum Telephone Company
Wm. H. and Joan A. Boatman
William R. and Dianne R. Boatsman
William "Billy" Wuorinen

RV "Duffy" and Ellen Larson, Larson Trucking, 1982. Photo from Caron Larson

This project would not have been possible without the generous help of donors and supporters who have made donations, loaned photographs, shared family stories, acted as consultants, provided tours, and helped in so many ways. We are deeply grateful and regret any mistakes made in the many details in the book.

It is not possible to recognize every logger. However, we hope the faces included reflect the spirit and hard work of every logger who ever tramped the Willapa Hills.

Karen Bertroch and Bryan Penttila, Co-Editors

When Logging Was Logging

100 Years of Big Timber in Southwest Washington

Co-Editors: Bryan Penttila and Karen Bertroch

Written by Karen Bertroch, Donna Gatens-Klint, Jim LeMonds, and Bryan Penttila

Introduction by Myron Metcalf and Epilogue by Brian Blake

An Appelo Archives Center Project

THE
DONNING COMPANY
PUBLISHERS

DEDICATION

I dedicate my work on this project to my son, Dude Bertroch, seen here working in the upper Salmon Creek area operating an excavator for Torppa Construction. Dude worked for Bob Torppa in the summer of 2010, then decided to move to Grays River. Torppa Construction is a good example of the businesses that the timber industry supports. They maintain roads and culverts as a result of the state's RMAP (Road Mandated Abandonment and Management Plan). It is my hope that Dude, as a second-generation "immigrant," will serve as an example to other young adults who may be considering a move to the area.

Karen Bertroch, first-generation "immigrant" in 2003

Photo by Greg See

Half title photograph: Photo of Norman Penttila by Bryan Penttila

The Donning Company Publishers
184 Business Park Drive, Suite 206
Virginia Beach, VA 23462

Steve Mull, General Manager
Barbara Buchanan, Office Manager
Pamela Koch, Senior Editor
Chad H. Casey, Graphic Designer
Priscilla Odango, Imaging Artist
Tonya Washam, Marketing Specialist
Pamela Engelhard, Marketing Advisor

Cathleen Norman, Project Director

Library of Congress Cataloging-in-Publication Data

When logging was logging : 100 years of big timber in southwest Washington / co-editors, Bryan Penttila and Karen Bertroch ; written by Karen Bertroch ... [et al.] ; introduction by Myron Metcalf and epilogue by Brian Blake.
p. cm.
Includes index.
ISBN 978-1-57864-705-7 (hard cover : alk. paper)
1. Logging—Washington (State)—History. 2. Loggers—Washington (State)—History. I. Penttila, Bryan. II. Bertroch, Karen. III. Metcalf, Myron.
HD9757.W2W46 2011
338.1'7498097978--dc23

2011034346

CONTENTS

Above: "Pockethold" with unidentified loggers.
Photo by Jerry DeBriae

INTRODUCTION

By Myron Metcalf

When I was growing up along the banks of Salmon Creek, on the Pacific-Wahkiakum County line, my parent's kitchen table would draw a variety of neighbors, friends, relatives, and "characters" who would weave wonderful word-pictures of their adventures in these rain forest hills. Stories of hunting outings and camping trips, of fishing days and the running of their trap lines . . . and of logging. Always stories of logging. And I would hear of the exotic places you will read about on these pages. Places where the logs came from: Cedar Creek, Savage Creek, down in Bear River, over on the Nemah, up above The Gorge and way up the Elochoman. Such great sounding and exciting and seemingly faraway places.

One of our regular visitors was an old bachelor logger who would tread a half-mile along a dark forest trail with his little two-cell Ray-O-Vac flashlight that sometimes worked and sometimes didn't. He'd come in, hump up with his back against the wood stove for a little while, and then settle at the kitchen table. He'd dump a healthy dollop of cream in his coffee, add a couple of shovels of sugar, and then proceed to pour it into his saucer for drinking. That's what saucers were for in this logger's raising. Bill was a yarder engineer . . . a "donkey puncher."

He had started his logging days around the turn of the century as a ten-year-old boy, greasing the skids in front of an ox-team when the first settlers started letting daylight into the swamp so they could grow spuds, peas, cows, hay, and kids. He had logged these hills during the ox-team days of Simon Benson, B. F. Brock, Louie Saldern, and the Brix Brothers. And he had seen the huge, isolated logging camps with the great logging railroad engines and the magnificent steam-powered yarders and loaders of the big operators like Crown Willamette, Deep River Logging Company, and Long-Bell. Now, in the 1950s, Bill was at the throttle of a powerful diesel yarder, a forerunner of those that rumble today.

The kitchen table was also where great and sometimes boisterous young men gathered (more boisterous around the Fourth of July and Christmas logging shutdowns) to tell their logging stories of the day. They were a group that had no doubt that with a good enough tail hold, enough horsepower and enough snoose, there wasn't anything between Chinook and Chehalis that they couldn't log. They were a go-ahead group of loggers' sons and other country boys that Uncle Sam had shipped halfway around the world to fight and win a war.

In winning that war, they learned how to do incredible things, with simple tools and machines, under the most horrible conditions. Boys who had plowed fields only with a horse suddenly learned how to build runways and roads with a Caterpillar. Others who had maybe driven a car a half-dozen times learned how to fly an airplane. And in some parts of the world, they logged and built sawmills. They learned to improvise, make do, patch up, and always keep moving ahead. And those that came back to these logging hills brought that same "git-er'done" attitude with them.

These were the men that built the truck roads and logged these hills for Weyerhaeuser, Crown Zellerbach, Grays River Log, Valley 8, Durrah-Martin, Reuben Penttila, and other gyppo loggers and private landowners. They logged with gasoline and diesel power, slowly moving from sled-mounted to mobile yarders. They advanced from cable shovel loaders to faster, more versatile hydraulic machines. The spar trees and the truck reaches of wood were replaced by steel tubes. They had moved from handsaws to two-man, and then one-man chainsaws by the 1950s, with continual improvements in power, reliability, and size. The more gentle slopes that the logging railroads could climb were left behind as these loggers moved farther and farther into these hills with truck roads that the engineers designed to get the timber in "the back end."

Myron Metcalf

Finally, the back end was reached . . . Dell Creek hooked up with Bear River; Grays River connected its logging with the upper reaches of the Elochoman, and the upper Naselle was just a short log truck haul from the Nemah. Then these loggers looked around and realized it was time to start all over again, in second-growth and third-growth timber with a new generation of loggers.

The new breed is at the controls of a new generation of high-cost machinery filled with electronics, computers, and read-out monitors. They get an instant read of diameters, lengths, and associated bucking instructions that will maximize the dollar component of each tree. The likes of Jerry DeBraie, C&C Logging, Willapa Logging, and Johnson Forestry are again moving logs where the ox teams plodded along 130 years ago. And with a zip of a machine-mounted saw, another tree is laid down next to an old growth stump, "three boards high."

The machines are faster, the log sizes and the logging crews are smaller, but the rain is still just as wet and the hills are still just as steep and muddy and the widow-makers are still just as deadly . . . and we still need to get the logs to town. And all of that is what makes this logging business so much fun. So pull up a chair to the kitchen table, and perhaps pour your coffee into a saucer, and enjoy this wonderful history of logging in southwest Washington.

Myron Metcalf, Forester
Winlock, WA

CHAPTER 1

Deep River Logging Company

by Bryan Penttila

When Bill Meserve and Adolph Olson got off the steamboat in Deep River, they had a clear goal in mind: to find the best logging show anywhere. The year was 1899, and their search had come at the behest of Simon Benson, one of the region's leading loggers and soon to become one of Oregon's leading businessmen and most celebrated philanthropists. Meserve and Olson had traversed both sides of the lower Columbia River looking for the right location where the timber was thick, accessible, and, most importantly, available at a reasonable price. The hills back of Deep River seemed like just the spot.

Standing tall and straight and for miles in every direction was one of the finest coniferous forests anywhere on Earth. In the bottomlands grew dense stands of Sitka spruce, hemlock, alder, and a jumble of smaller growth. Just beyond, Douglas fir stretched skyward to astonishing heights, while spire-topped western red cedar, some near twenty feet in diameter, sprang from the rain-leached soil. Here and there charred snags and young regrowth offered clues that a catastrophic wildfire had razed large swaths around Grays Bay in the 1830s. Windstorms and advanced age, too, had left their telltale marks, evidenced by the downed and decaying trunks scattered about. It was all a wonderful tangle, this old growth forest, and for Simon Benson, the starting point for another great logging operation.

Simon Benson epitomized the American dream. He came into the world as Simon Bergerson, September 9, 1851, on a small farm in Norway. Immigrating to the United States in 1867, he began working on farms and in logging camps in Wisconsin before opening a general store in 1876. Three years later the store and all his merchandise burnt, leaving Bergerson—now Benson—near broke and in need of a new start. He collected his belongings and with his young family moved west. The Bensons landed in Portland where Simon went to work in the woods. He kept a close watch on his earnings and soon bought a parcel of timberland. With six unbroken steers and one employee, he began to carve away at his new holdings. In this fashion Benson became acquainted with the hardships and rewards of a pioneer timberman, and, by the late 1880s, he had a small crew rolling logs into Beaver Creek, a stream between Rainier and Clatskanie.

It was here in the hills around Rainier that the various strands of Deep River Logging Company first began to intertwine. Provisioning Benson's camp on Beaver Creek and extending him credit when he had little else was the Portland wholesaling firm of Everding & Farrell. The partnership dates back to 1867, when

Henry Everding and Sylvester Farrell established a commission house in a little one-story brick building on Portland's Front Street. Everding & Farrell's business grew around selling grain, canning salmon, and feeding loggers. Over the years the company became financially involved in some of these timber operations, partly to safeguard their interests but mostly because they were profitable investments. Such would be the case in Deep River.

Olson Brothers' bull team logging show and ox barn, probably near Westport, Oregon, in the 1890s. France Olson leans against one of the oxen with his wife, Lottie, at his side.

Back in 1889, one of the loggers in Benson's camp filling up on Everding & Farrell's beef and beans was William Nelson Meserve, better known as Bill. His parents, Dr. John and Sarah Meserve, lived close by at a fledgling settlement they had named Delena. Just over the hill from Delena, on the Rainier side, Bill Meserve had gotten his first taste of logging working for the Swedish-born Olson brothers—

August Olson, logging manager and part owner of Deep River Logging Company from 1900 until 1926.

Charles, France, Adolph, and later, August. The older siblings had immigrated to America separately during the 1880s and, by the end of the decade, were running a logging camp near Rainier. In 1890, their youngest brother, August, joined them.

The Olsons first hired Bill Meserve, then a teenager fresh from the prairies of Nebraska, to work as a *skid greaser*, slathering liquefied tallow on the cross skids that made up the bull team logger's great thoroughfare called a *skidroad*. France was the teamster in charge of the oxen, a position of high esteem that in the western woods became known as the *bullwhacker*. Adolph handled the ropes and blocks and was responsible for getting the logs from the stump to the skidroad. His title was that of *hook tender*. "Frans was an expert ox driver," Meserve recalled a half-century later, "and handled a five-yoke team so well that with Adolph on the rigging and Frans back of the oxen, they always got out the logs."

Meanwhile, Simon Benson's efforts in the Oregon fir had proved successful. In 1890, he secured the cutting rights to four thousand acres on the Washington shore behind Cathlamet. Benson scrounged up some secondhand steel and laid a narrow-gauge railroad up Birnie Creek and into the stand. That was in 1891, and the following year he replaced his oxen with a steam-powered donkey engine. Benson was not the first to introduce steam power to the woods of the Pacific Northwest, as some later chroniclers would claim, but he was certainly among the first to realize the full-scope profits to be made from the transition. In 1894, he located a rich tract of fir about a dozen miles upriver at Oak Point and soon had loggers at work there. Recognizing good talent, Benson hired August Olson to run one of his new camps. Oak Point proved to be a lucrative show, allowing Benson to buy out the Portland lumbermen that had backed his venture. By 1899, "The Boss Logger of the Columbia" commanded twelve miles of railroad, ten steam donkeys, and five locomotives.

With a mind toward expansion, in 1899 Benson set his sights on Deep River. The notion to build a railroad from Deep River into the big timber country of the Salmon Creek valley had been on lumbermen's minds long before Simon Benson took an interest in the area. But before 1900 there had always been more accessible timber that could be brought to the Columbia River at a lower cost. Those stands were running

mighty scarce though, and once it was decided that Deep River would be the next operation of the Benson Logging & Lumber Company, the Boss Logger sent Bill Meserve out to buy up as much timberland as he could.

According to Meserve's field notes from that time, most homesteaders' claims—usually 160-acre "quarter sections"—contained between four and five million board feet of standing timber. Meserve worked tirelessly throughout the spring of 1900, traveling from homestead to homestead, meeting the occupants, cruising their timber, and offering them terms. In addition to "taking the option" on settlers' claims, Benson purchased entire square-mile sections—each 640 acres—from the Northern Pacific Railroad Company. By May, a local newspaper reported that Benson possessed nearly twenty square miles of timberland north of Deep River, most of it Township

Nearly seventy Deep River loggers pose for this picture, circa 1903. In the back row, near center, France Olson stands with his arms crossed. To his right is his wife, Lottie, and between them, their son Burnell.

A trainload of logs at the dump sometime between 1909 and 1912. The large building in the background is Deep River Logging Company's sawmill, and beyond that, the roofline and drawbridge in the town of Deep River.

11 North, Range 8 West. With his work at Deep River done, Meserve settled in nearby Grays River where he became a successful merchant and later served a number of terms in the Washington state legislature.

In May of 1900, Simon Benson, Everding & Farrell, and Adolph and August Olson signed the paperwork creating the Deep River Logging Company.

Railroad building commenced at once. Laborers began clearing and grading the right-of-way up the West Fork of Deep River following a route surveyed by Bill Meserve. The path was first *swamped out* with dynamite and a donkey engine before *Fresno scrapers* pulled by horse teams began to move dirt. By the fall of 1900, crews had put in about two miles of railroad, sufficient trackage to reach the company's holdings in Salmon Creek. The exact cost of railroad construction is not of record, but a contemporary estimate placed the expenditure at over $50,000—a mighty expensive undertaking for the time. To house the sixty or so loggers, a camp was put up not far from where the Deep River Mainline crosses Salmon Creek Road today. The boys went to work, and in October the first trainload of logs came rumbling off the hill behind a twenty-ton side-winding Shay locomotive.

Despite his key role in getting the Deep River operation going, Simon Benson sold his interest in the venture after only three or four years. He never indicated why he divested with his Deep River stock, but it appears that his shares were sold to Richard Everding, a second-generation member of the Everding & Farrell firm. Benson would go on to even greater fame and riches, first by opening up the fir country behind Clatskanie with an expansive railroad. He then took a gamble and built a sawmill way down in San Diego where he correctly predicted a big demand for lumber. To get his logs down the coast, Benson began making gigantic, cigar-shaped log rafts, some over one thousand feet long. In the years to follow, Benson sold most of his lumber-related investments, became even wealthier dabbling in Portland real estate, built the Benson Hotel, and dedicated his retirement to philanthropy.

Completed in 1911, Deep River's tunnel eliminated the adverse grade over Mikkola Pass, which separated the Deep River and Salmon Creek valleys. Logs were hauled through the tunnel until the operation ceased in 1955, after which the portal was imploded with dynamite.

Even before Simon Benson left Deep River Logging Company, the day-to-day management of the firm had been the responsibility of the Olson brothers Adolph and August. Everding & Farrell may have held a majority of the stock, but for some three decades, the logging operation was the brothers' domain. Their influence in the woods was absolute. To local residents and employees alike,

Deep River Logging Company and the Olson brothers meant the same thing. Adolph settled into the role of superintendent while August managed the workings in the woods. Not surprisingly, their brothers came to Deep River too. France became a foreman and Charlie a masterful steam donkey engineer. A lifetime of hard work hastened France's demise—he died in 1911—while Charlie seems to have led a rather unassuming existence before retiring around 1920 and living out his years in August's home in Portland.

At Deep River, the largest hindrance facing the Olson brothers was the railroad pass into the Salmon Creek valley. For a full decade, it tormented the operation. Despite continual improvements, this steep, swerving stretch of track was prone to mudslides and the cause of several significant train runaways, derailments, and collisions. The worst incident occurred in the fall of 1904, when two trains crashed into each other on a blind corner not far from the pass. This accident took the life of eleven-year-old Frank Heiklin, a local boy who was hitching a ride back to his family's farm, and was the basis for a costly lawsuit filed by a traveling salesman who was also injured.

If the logs were to be hauled out of the Salmon Creek valley for years to come, something would have to be done about this restriction. The brothers decided to do something almost unknown in the big woods—they would build a tunnel. In the spring of 1909, the company put up a small sawmill near its log dump and began cutting brace timbers and ties. With shovels and picks, workers began excavating the seven-hundred-foot-long portal that was to be sixteen feet wide and twenty-one feet tall. They filled small dump carts with earth, which were wheeled over narrow gauge tracks to a nearby gully and dumped. Inside the tunnel, the monotony of the work was periodically interrupted when gas pockets erupted into flame. Slowly, through the mud and siltstone, the tunnel began to take shape.

In 1911, the first train rolled through the hole in the hill. The new route—with its 3-percent grade—enabled Deep River to supplement its slow, hard-pulling Shay locomotives with faster direct-connect rod locomotives to serve its increasingly distant logging works. Though the tunnel was a significant improvement over the old pass, it was not without problems of its own. The *Skamokawa Eagle* reported in 1916 that the portal had cost $54,000 to build and to that date had required an unthinkable twelve million board feet of brace timbers. Despite the veritable forest of wood within the tunnel, an occasional pocket of mud would ooze through the walls and onto the tracks. For this reason, the Olsons thought it prudent to hire a watchman to patrol the tunnel. Three Hermans—Herman Cragell, Herman Hyrkas, and Herman Anderson—filled this position over the years.

During the period of tunnel construction, August Olson married Grace Hall of Beaverton, Oregon. He may have become acquainted with Grace through her five brothers—Bill, Jack, Martin, Andrew, and Mike—all of whom worked for Deep River a good many years. Martin Hall, in fact, went to work for Deep River in 1904 and did not retire until he was diagnosed with terminal cancer a few months before the company shut down in 1955. Other members of the extended Hall clan were on the company's payroll too, including Tom Kennedy and John McConnell.

Most of the Olsons' employees came from Scandinavia. A number of these immigrants, so the story is told, were guided across the country by shipping instructions pinned to their lapel that simply read: "Olson Bros. Camp, Deep River, Wash." Many of these new arrivals came from the Olsons' native province of Varmland, in western Sweden, including John and Carl Bergquist; John and Gunnar Haglund; Axel and Ernest Larson; Peter Magnuson; August, Oscar, and Gunnar Nelson; Bertell "Swede" Pearson; and Helge Peterson. Finns, too, represented a large percentage of the Deep River logging crew. Among their ranks stood foremen Emil Ruitta and Toivo Hill.

Loading logs with an "A-frame" and "end hooks," probably before the First World War. On the log, left to right: Helge Peterson, Ole Bjork, and Gus Magnuson.

Around the time Deep River's tunnel went into service, the company installed a traveling unloader to replace hand-powered screw jacks in rolling the big sticks off the rail cars. Fred Hendrickson, who lived on a nearby farm, operated the unloader from 1912 until the dump was relocated nearer the town of Deep River in 1929.

When the tunnel opened, Deep River had two camps running full blast and employing more than two hundred loggers. One camp was located at a place later called Busse Canyon. The company's loggers had been working in that vicinity since 1905 when three miles of steel was laid into the upper Naselle River valley with spur tracks tapping Dog Salmon Creek among other areas. This camp closed around 1912. The other camp was located along Salmon Creek where the company's mainline was pushing ever deeper into the tall fir. From its original location, the camp had been moved upstream near the Rinell family farm around 1903 and, about four years later, relocated still farther up the valley hard by the Haapakangas (later Catlin) farm.

By 1914, Deep River's camp was on the move once again, following the timberline up the Salmon Creek valley. The bunkhouses and other portable buildings were slipped onto railcars and ferried up the line a couple of miles where they were placed along the tracks. The new site was a pleasant location, a broad logged-over flat at the base of a gentle hill, flanked by the lazy meanders of the creek. This would be the fourth camp operated by the Olsons, and the longest lived. It would be known officially as Camp 1, but to its residents it was simply called Toonerville. The camp's informal title came from a popular comic strip of the time called "Toonerville Trolley." It was the handiwork of cartoonist Fontaine Fox and featured a ramshackle little streetcar and a cast of good-hearted but eccentric characters. Either by design or luck, the Olson brothers attracted a loyal corps of *homeguard* loggers, many of whom spent their entire careers with Deep River.

From Toonerville, Deep River's loggers ranged over the hills to the east and left a swath of stumps all the way to the head of Hull Creek. They pushed northward, too. One branch of the railroad climbed the face of "The Big Hill" while the mainline sliced through "The Big Cut" and emerged above the headwaters of the Naselle's East Fork. There, the company's engineers found a broad flat, too expansive to be reached by conventional logging methods. They decided to build an incline. From the mainline, a spur track was laid down the contour of the 12-percent hillside. Used between 1925 and 1926, the incline allowed empty cars to be lowered fifteen hundred feet down into the basin and, once loaded, pulled back up to the mainline by a donkey engine. Though the incline worked slick, it pointed to the increasingly difficult terrain that the logging crews were headed into.

A tremendous amount of time and money went into building a logging railroad. To keep grades consistent in hilly country required a great deal of trestle building. This Deep River crew has just capped the third "bent" on the trestle they have just begun.

After a career of bone-tiring work, August Olson sold his shares in Deep River Logging Company and retired to his farm near Salem, Oregon, in 1926. By this time, Adolph's son, Herbert, was also helping to manage the firm. Herbert grasped the executive end of the business well, but, according to some old-time loggers, he often came up short on practical logging matters. Just the same, in keeping pace with the industry, Adolph and Herbert decided to try a new logging method that promised to increase production.

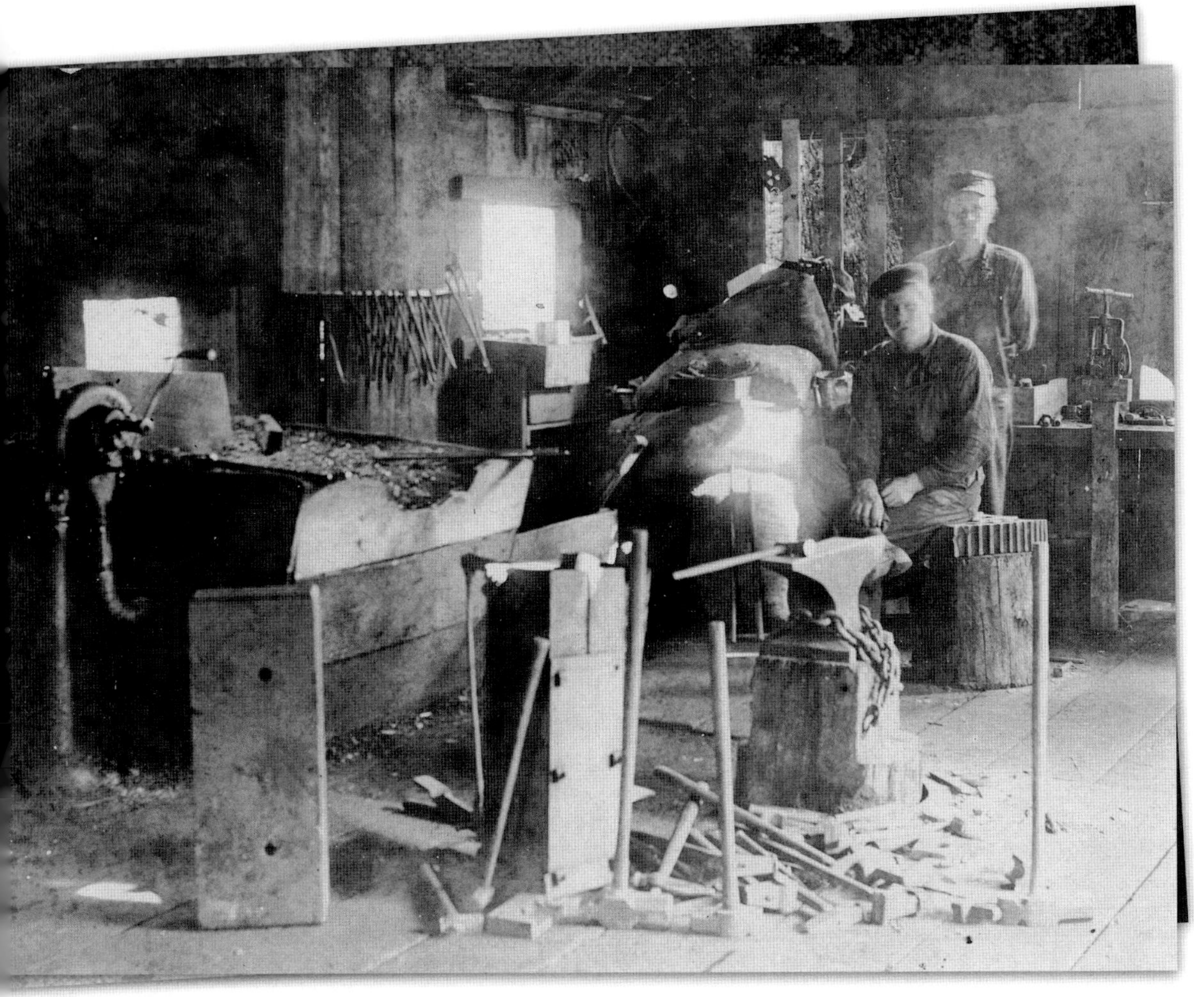

Blacksmiths were among the most critical personnel in every early logging camp. Here, brothers Albert and Victor Harpet take a well-deserved break from the anvil.

In 1926, Deep River sent one of its steam donkeys to Seattle to be reconfigured for slackline logging. The slackline system employed a heavy steel *skyline* cable and *carriage* that was capable of snagging logs from several thousand feet away and suspending them in the air as they were reeled in. It represented a relatively new and increasingly necessary method for the Olsons. Then as now, the upper East Fork country is steep, rugged ground. Deep River's railroad followed the stream bottom, posing the company's loggers with the challenge of getting the oversized cuts of fir off hillsides that rose some fifteen hundred feet above the rail line—a mighty formidable task.

Unfortunately, Deep River's homeguard loggers did not take to slacklining. The production goals that the Olsons had expected from the reworked machine were not being met, so they contacted the manufacturer, Washington Iron Works, for help. Washington sent Herbert Olson and foreman Toivo Hill on a tour to Puget Sound to see Buckley Logging Company's slackline machines yarding logs without a hitch. They returned with new ideas, but Deep River's slackline crew still did not produce. Once again, the Olsons called upon Washington Iron Works for assistance. This time, they sent Stewart Hood, a well-versed veteran of the Buckley woods, down to Deep River to troubleshoot in the fall of 1929.

Hood had seen slacklines churn out wood on some of the most rugged shows around Puget Sound and could only guess at what awaited him at Deep River. "I came down and old man [Adolph] Olson wanted

me to go out and look where they wanted to use it," recalled Hood. At the time, Deep River's slackline was set up at the confluence of the East Fork and the Little East Fork. "It was the most beautiful setting—something you might dream of. At least for a guy who had skyline logging in his head." Olson asked Hood how many logs he could get with the yarder in a day's time. None, replied Hood, without a crew that was up to it. Olson shuffled his homeguard loggers to other sides and brought in a few new highball men selected by Hood.

Deep River's first Baldwin No. 7, a 2-6-2, circa 1925, tends to a landing at the site where the company's Bachelor Camp would later take shape. Note the homemade crew car sitting to the right of the locomotive. Photo courtesy of John Bergquist

William "Wild Bill" Allison, a gung-ho logger, became hook tender. He quickly drove off Gunnar Nelson, the rigging slinger, who was replaced by Johnny Ross. Chagrined, Nelson and another Swede, Gunnar Haglund, chose to work on one of the company's high-lead sides. This created quite a bit of friction in camp, but the slackline's landing was soon stacked high with wood. Some of the younger locals fit in well with the new regime, Hood remembered. Otto and Charlie Kieski proved

Shay locomotive No. 3000 heads across the one-hundred-foot-tall Slop Canyon trestle with a load made up of cuts from a single fir tree. The bridge spanned the East Fork of the Naselle River just below Bachelor Camp.

worthy as third riggers, working in the back end of the operation, and Dave Kingelin and Victor Wuorinen fit right in as choker setters. Unfortunately, just weeks after getting the slackline side running, Ole Bjork, a Deep River veteran, was crushed beneath a fir log while loading railcars under the spar tree.

By 1931, Adolph Olson was ready to retire. The Great Depression had crippled the logging industry and removed nearly all of the profit from the Deep River operation. Already well into his sixties, Adolph wanted to focus more on his golf game and less on a sagging business that was nearing the end of its timber supply. Herbert chose not to continue the family business and opened an industrial supply store in Portland. On June 4, 1931, Adolph and the representatives of Everding & Farrell sold the Deep River Logging Company. Adolph passed away eight years later in Portland. His younger brother, August, had preceded him in death two years earlier. In the thirty-one years that the Olsons ran the operation, no less than a billion board feet of logs hit the water at Deep River—a mighty impressive record by any measure.

Loading logs with a "hayrack boom" in the early 1930s.

The new Deep River Logging Company was composed of the Koster Products Company, the Knappton Towboat Company, and Norton Cowden, each holding a one-third share. Orville Miller, who had run Koster Products' operation back of Astoria, and Cowden, whose family owned the Silverton Lumber Company, in Silverton, Oregon, became the managers. Cousins Walter and Herbert Brix, associates with the towboat company, became foremen. Facing the same dire economic conditions that had confronted the Olsons, the new firm made some difficult choices. For months on end, the camp was closed until the log market improved. They let go or demoted long-term employees in favor of loggers who had worked for them elsewhere. It was a painful affair for all the old Toonerville homeguard but certainly not an uncommon occurrence during the Great Depression.

Deep River's No. 7 locomotive steams past the cookhouse at the Naselle Camp en route to the dump in 1950.

By 1932, logs were once again rolling down the Deep River line. The new firm had secured some three thousand acres of timber at the headwaters of Alder Creek, a tributary of the Naselle River, from the Werley Timber Company of South Bend. Loggers once again filled Headquarters Camp, a facility that had been built three miles up the railroad from Toonerville in 1928. This is where all of the single men stayed during the workweek—hence its nickname, "Bachelor Camp"—while the family housing, locomotive barn, and machine shops remained in Toonerville. During this time, a succession of legendary foremen came and went through the Deep River operation, including Earl Hitchman, Clarence Vanderjack, and Mike Kauffmann.

Thrift was standard operating procedure at both of Deep River's camps during the Great Depression. In the early 1930s, both the *West Coast Lumberman* and *The Timberman* magazines ran feature articles that described the operation as having "no

In 1939, Deep River began hauling its logs out of the hills and to the railhead on motor trucks. Earl Spargo's small fleet of big rigs did the trucking. Today, the cement slab from his truck shop is the last visible reminder of Deep River's Naselle Camp.

frills." It was jokingly said that Norton Cowden even discouraged his loggers from walking on the railroad ties with their caulk boots so as not to chew up the wood. To keep its aging machinery and rolling stock in top-notch working order, Deep River was blessed with two Swedish-born mechanical geniuses, Freddy Strom and Eric Johnson. With their ability to fix and maintain the company's hardware, Deep River saved untold amounts of money during the Great Depression and after.

In 1932, Deep River made a rare Depression-era expenditure, purchasing a Caterpillar Sixty crawler tractor with a bulldozer unit. As the story is told, a throng of old-school loggers stood looking in abhorrence at the newly arrived dozer as it sat on the barge moored in front of Pentti's Pool Hall in Deep River. From the crowd of disapproving onlookers stepped young Karl Kruse, who volunteered to navigate the new fangled contraption onto dry land. For his initiative, Kruse was given the job of cat skinner, at which he proved to be quite proficient.

Deep River's No. 7, better known as "Skookum," was among the most famous logging locomotives in the Pacific Northwest, in part for its unique 2-4-4-2 wheel arrangement. Built in 1909, the engine arrived at Deep River around 1933 and was the workhorse of the operation until it derailed in 1955.

As the Werley timber neared exhaustion, it looked as if the old firm might be through. But, in the spring of 1936, Orville Miller, C. L. Koster, and Norton and R. A. Cowden joined with Portland sawmill executive O. J. Evenson and his four sons—Willard, Clarence, Justin, and Franklin—to incorporate the Deep River Timber Company. The firm was organized to purchase a large tract of Weyerhaeuser timber, some eight hundred million board feet of it, lying along the upper Naselle River. Deep River Logging Company would harvest the new concern's timber, but first it would have to relocate. In 1938, a new rail line was built into the Naselle valley, and shortly thereafter, the portable buildings from Headquarters Camp and Toonerville were transferred to the new site.

Deep River's engineer, "Ten Percent" John Forsberg, designed the new operation as a railroad logging show. He and his crews laid out miles of right-of-way through the tall timber along the upper Naselle. The plans for an extensive rail line were scrapped, however, in favor of hauling the logs out of the hills with trucks. Under the supervision of logging superintendent C. J. "Bud" Underwood, two transfer stations, or reloads, were built near the new camp where logs would be taken off the trucks and put on rail cars for delivery to tidewater. Earl Spargo's General Hauling Company got the contract to haul the logs from the woods to the reload. Though in the early years Spargo's fleet was in a questionable state of repair, in the eighteen years that he hauled for Deep River, only one fatal trucking accident occurred, that being in 1950 when a log slid onto the cab of a truck crushing the driver, nineteen-year-old Robert Olmstead.

The shop crew at the Naselle Camp in the early 1950s. Left to right: Master mechanic Freddy Strom, blacksmith Eric Johnson, Bruno Postatni, and Frank Guenkel.

To speed the harvest of its vast new holdings, Deep River began assigning logging contracts. Smith Creek Logging Company, a co-op formed by thirteen Finnish-Swedes from the Raymond-South Bend area, built a camp on Lane Creek, near Naselle, and got to logging. From 1937 until 1945, "The Swedes" cut and hauled the big round stuff off the western portion of Deep River's acreage. Another of Deep River's contractors was Bryce "Scotty" Caldwell. A native of Quebec, Caldwell came west during the Great Depression and by 1936 was building railroad trestles for Deep River. He soon expanded into gyppo logging and, across the Naselle River from Deep River's new camp, began hauling logs to one of Deep River's reloads. Eddie Hurse, one of Caldwell's most skilled cat skinners, became a contact logger for Deep River after Caldwell pulled up stakes in 1949.

In the meantime, Deep River's loggers were busy in the hills flanking the North Fork of the Naselle. After the Second World War, the camp housed nearly 150 men. Every day, a bus from Longview would bring new workers, picking up those who could not make it as loggers. Assisting Superintendent Underwood in managing the ever-shifting crews was a Skamokawa-born logger, Art Engelson. Art married a local girl, had several children, and was proving to be a popular and efficient boss of the operation. Then, as he was getting ready for work early one Monday morning in the fall of 1954, the forty-four-year-old Engelson succumbed to a heart attack.

Trainmen Stewart Hood and Dewey Stinchcomb inspect the "Skookum" as Assistant Logging Superintendent Art Engelson looks on from his pickup truck. Photo taken just below the "reload" at the Naselle Camp, March 1950. Photo from Evelyn Engelson Collection

A few months later, a less dire though more symbolic event occurred on the railroad tracks just below camp. While bringing a train of empty cars back into the woods, Deep River's famous 125-ton Baldwin locomotive, No. 7, hit a loose rail and pitched over at a dreadful angle. Engineer Stewart Hood and Fireman Dewey Stinchcomb escaped the teetering engine and watched as, a few minutes later, the iron horse lay over on her side with a terrific thud. Though Deep River still had a couple of locomotives on hand, when the big engine tipped over, it marked the end of an era. It was no secret that the firm's timber supply was running awfully thin, but after some two generations of logging, it was hard for locals to believe that the show would be coming to an end.

In September of 1955, the last trainload of logs pulled out of the Naselle camp bound for Deep River. It headed downriver, past Busse Canyon where the company's loggers had toiled a half-century before and clickety-clacked by No. 7, still prostrate along the tracks. The train crossed the Naselle River a couple of times and began its long, gentle climb over Hungry Highway hill before its descent into the Salmon Creek valley. There it steamed over the long trestle spanning the Hyrkas farm where Deep River had its first camp back in 1900. Then down through the tunnel—the most notable landmark on the line—and into the town of Deep River where the logs met tidewater. The work of dismantling the camp, tearing up the rails, and disposing of the equipment kept a skeleton crew busy until the end of the year when the company was dissolved.

Bill Meserve did not live to see the end of the operation that he helped locate. He passed away in 1953 at the ripe age of eighty-six. Meserve and Olson certainly fulfilled Simon Benson's wish to find the best logging show anywhere, and for evidence one need only look at Deep River's fifty-five years of logging. Today, little remains of this long-lived enterprise. The great, weathered stumps of the forest that Deep River dismantled are the most unmistakable reminder of the company's exploits. Newer generations of loggers have harvested a second and even third timber crop from the very same hillsides, yet the old stumps remain, reminders of one of the best logging shows of all time.

CHAPTER 2

Growing Up in Toonerville: An Essay

by Karen Bertroch

From the newspaper *Grays River Builder*, May 6, 1937, "Deep River Log Camp Has Real Pioneer Spirit."

> *Truly, these women have the real spirit of the pioneer as forsaking the easy ways of modern life they follow their men to the primitive surroundings of the logging camp. As I went from house to house I could but marvel at the wonders some of these women have accomplished. The houses for the most part are flimsy shacks but dainty curtains at the windows, the artistic feminine touch of a little wall paper, a little linoleum, shining cook stoves, tell the story of women who, realizing the inexpediency of going to great expense, have contrived to make a home that is attractive and comfortable with minimum expense.*

Unidentified man with child in a logging camp. Photo from Archives Collection

In today's world they would be called shacks, but to the folks who lived in them, they were simply called home. Built of rough lumber, they were small, sturdy, two-room dwellings with an outhouse out back hanging over the creek. A hand pump in the kitchen provided water, and a wood stove supplied heat and a place to cook. There were, however, always curtains at the windows. Women made the flimsy shacks into homes as best they could. The front door faced the tracks with only a few feet separating the house from the trains that ran by several times a day. Sometimes there was room for a small yard, but not always. As primitive as it may seem, to the children who grew up in early logging camps, this was home.

The late 1800s were tough years. Ted Swanson Sr. described those years in an interview in 1975: "I was nine years old when the Depression in 1893 came about. As I remember it, there wasn't any money around. I lived out in the country, of course, but it was the same all over the United States I'm sure.

Four unidentified children at Olson Bros. Logging Camp. Photo from Archives Collection

Nobody had any money. There was no work, and my folks had two men that came into our house, and they had no job; so my folks, they let 'em live there with us and helped a little bit once in a while in doing a little work to help pay."

Company camps came about in the early 1900s in response to the logging railroads being built to bring logs to the waters of Deep River, Skamokawa Creek, and the Elochoman River that run into the Columbia River, and the Naselle River that runs into Willapa Bay.

"Tramp loggers" in the early 1900s moved from camp to camp much as the American cowboy moved around ranch "outfits" in the late 1800s. Tramp loggers were housed in bunkhouses in the working months, but when winter set in and work stopped, they moved on to other areas seeking work. Early in the 1900s, employers realized that, if women were involved, men would tend to stay, so family houses were built to accommodate the women cooks, women kitchen flunkies, teachers for a camp school, and wives of the woods bosses and trainmen so essential to the camp's success.

Mr. and Mrs. Emil Bjork on the porch of their company house at Olson Bros. Logging Camp.

"Companies needed to insure [*sic*] a more stable labor force near the logging sites. Early ideas about catering to married workers later found fulfillment in the towns built between 1910–1930 as the mill owners, plagued by strikes, labor unrest, and gross employee turnover, began providing housing units for families" (Kenneth Erickson, "The Morphology of Lumber Settlements in Western Oregon and Washington" [Ph.D. diss., Univ. of California, Berkeley, 1965]).

An early camp, the Saldern Logging Camp in Grays River, started in 1899. Marie Klint Fauver, a local historian, has lived in the Saldern Camp's superintendent's house at the bottom of KM Hill in Grays River all of her eighty years. Marie remembers that when she was a child, "The air was blue in the fall because the sky was full of smoke. Everyone was burning old trees and stumps. The smoke hung over the valley most of the winter. The camp was located at the bottom of the current KM Hill near Klint Creek. My dad said there were over 100 bunkhouses in the camp."

Logging jobs were plentiful in the early 1900s, but it was such hard work, it often drew immigrants who didn't qualify for "higher" employment. Many came to the Pacific Northwest in search of work in order to build new lives for themselves and their families. Many saved what they made to pay passage for family members back in their homelands so they could be together in their new home, America. They were willing to do whatever was required to start their new life. They brought skills, knowledge, and openness to learning, as well as strong traditions of faith, family, and community.

In a 1973 interview with Carlton Appelo, Toonerville teacher Lulu Hall tells of her three years teaching school from 1916 to 1919. "I had 7 pupils from 25 families. I taught first through eighth grades during the day, then I taught night school for adults who wanted to learn to read English so they could take their citizenship tests. Many couldn't read at all. I had Greeks, Italians, Finns, and then there sat the Italian section boss. I was only 19 at the time. There were over 200 men in the camp with bunkhouses for single men up above, and family camp just below. Four women worked in the cookhouse."

Marie Fauver at her home in 2011. Bertroch photo

Right: Lula Goodrich, teacher, and her students stand on the porch of the schoolhouse at the Olson Bros. Logging Camp, circa 1925. Photo from Archives Collection

Typical image of a mess hall at a logging camp. Some cooks fed as many as two hundred men at a time. Women worked as "flunkies" to help prepare the food and serve meals.

In the railroad camps, food was a required element that could make or break the reputation of the camps and the companies that ran them. Both women and men cooked and ran the cookhouses, but mostly women worked as kitchen flunkies helping prepare the huge amounts of food required for immense breakfasts, lunches, and dinners that kept loggers going over a long day's time. To keep the men going hard all day required up to five thousand calories a day.

Marlene Wirkkala Wiitala's grandparents owned the Shamrock Hotel in Deep River. She describes the meals her grandmother, Hulda, cooked. "Meals served were real 'logger meals.' Breakfast consisted of fruit, cold and hot cereals, bacon and eggs, hash browns, pancakes, toast, coffee, and milk. We provided lunches for the loggers, too. Lunches were two whole sandwiches, fruit, pastry, and a beverage. Dinner always included soup or salad, meat, potatoes, vegetables, fruit, and pastry. Diners were required to carry their plates to the kitchen or face a five-cent fine."

Margie Rinell Metcalf in 2011. Bertroch photo

Logging was dangerous work with little medical help available for accidents, although in 1914 Dr. Glen Prime had an office at Deep River's Shamrock Hotel. Not only loggers were victims of accidents. Margie Rinell Metcalf remembers that on Christmas Eve in 1922, "Father's youngest sister, Aune, came down the track in the dark on the speeder from Olson's Camp with other family members. The 3-wheel speeder hit something on the track nobody could see. She slid off and fell at least 20 feet into Salmon Creek. The lantern went, too. No one could see her, although we could hear her calling her husband's name. The next day her body was found downstream. She was in her twenties. She and her husband were newlyweds living at the camp."

Loggers with families didn't like to keep moving around from camp to camp, so they were happy to work where a home was provided as part of their pay. They could live there as long as they continued working for the company; thus, camps were truly the first "company towns" of the area. In the winters when work was shut down because of snow and bad weather, families lived rent-free in their homes and were grateful for the house and the community of others living in the same situation.

It was a hard life for many of the families. Margie Rinell Metcalf's father died in 1935, leaving her mother with four daughters and one son. "Mother supported us. She had a tough time. We kids made our own fun and never missed what other kids had. We helped Mom gather alder wood for heat, we peeled bark and picked berries."

In the Upper Salmon Creek area, there were two camps established by the Olson brothers, the lower camp and the upper camp. About twelve miles up Salmon Creek sat Camp #1, the lower family camp with twenty or so small houses built for families. Three miles further up at Bachelors Camp, the single men lived in bunkhouses. There was also a cookhouse and the train shop up at Bachelors Camp.

Over time, the lower family camp was named "Toonerville" after a popular comic strip in the newspapers called "The Toonerville Trolley." Established sometime around 1913–1915, the camp was a prominent

Claire Sundquist at sixteen years old breaks the ice off the family's waterwheel on the Sundquist farm just below Toonerville. Photo from Claire Sundquist McCollum Collection

ongoing community until it was closed in 1938, at the time the upper Naselle Camp was built. Although built in a long, narrow canyon, Toonerville was one of the last open areas, as the train tracks were built climbing up into the hills. Bachelors Camp was one of the last places where a camp could be built along the tracks. Just past that site, a large trestle connected wide spans between steep hills, so the two camps were built as high as possible. Camp houses were built to fit onto a railroad car so they could be easily moved from one place to another. Toonerville's houses, however, stayed in one place for more than twenty years.

Claire Sundquist McCollum spent her childhood in Toonerville, then on their farm at the low end of the camp. She remembers that children wandered in and out of each other's homes as if all the homes were their own. "Everyone knew everyone else. There were probably 20 kids or so in Toonerville."

Barbara Gonzales wrote stories about her childhood memories. "While most people think that children who grew up in logging camps had hard lives and were poor, here are some stories that show this is not the case in every circumstance.

"Mickey and Butchie Hurse's dad, Eddie, had his own business as a road contractor at the upper Naselle camp for part of the time that they lived there; he ran a Caterpillar. Mickey can remember going to work with his dad regularly. He recalls that once there was a hole on the plate deck and he got too close to it. His foot got sucked into the area between the drive shaft and the engine. There was so much pressure that it ripped his shoe right off, and they never found it. He stated, 'I went home without a shoe, but it could have been my foot that was missing.'

"Mickey and Butchie lived right next to the train tracks, and Mickey loved to play in the ditch when it filled with water. You could find him there with his boats. Fathers in camp began to realize that they needed to make a place for the children to play, so they cleared a spot and built a swing set, but it was only played on a couple of times. 'It wasn't that there was anything wrong with it,' Mickey said. The children just found other things to do. On Sundays the train was parked up at the camp, so the children played on it."

Evelyn Johnson told a story about Bunny Engleson that happened before they moved there. Bunny was about five or six, and after the speeder had finished its morning run, she got into the speeder and took off with it. The men saw the speeder moving and ran after it. "I don't know what kind of trouble she got into."

Bunny's brother, Eddie, used to repeat the bad language he heard from the loggers. He never got into

trouble saying them, but Bunny did. So Bunny and Evelyn would "go out somewhere, sit by a tree, and say all the bad words we could think of." The trees didn't seem to mind. Kids used to go to the cookhouse and get treats because Diane's mom and some of the other moms worked there. They all felt very safe in the camp and nobody locked their doors. The kids never bullied one another, but like all kids, they did pick on one another a little bit.

Shirley Olmstead thinks fondly of growing up in the camp. She remembers playing in the woods and in the river. The kids used to walk across a log to get to the other side of the river, knowing that if their parents found out, they would get in trouble. Playing on the train tracks was always great fun and putting pennies on them was a favorite thing to do. She loved to pick huckleberries and knew the best places to find them.

The children in camp grew up close, and those who remain still keep in touch with one another. They used to have a reunion every couple of years, but now most are gone. Even though their lives led them down different paths, they have a bond from the past that will always be there.

Claire Sundquist McCollum and Billy Wuorinen on Claire's porch at her home in Labam, Washington, in 2011. Bertroch photo

The families who lived there became a tight community and were very close. Even though many spoke the language of their native homeland, most tried to learn English so they could have one common language. Toonerville names reflected the international flavor of the community: Sanchez, Larson, Wuorinen, Rinell, Tienharra, Bjork, Anderson, Hyrkas, Bond, Sundquist.

In the winters when work was shut down and men had no income, the men hunted, the women fished, and everyone shared what they had. In the years there before a road was built connecting the camp to Naselle, game could be brought into camp openly, and those who tried their hand making moonshine were not worried about being caught. The tracks were full of snow, and no one went out or came in. In later years, if the game warden was seen headed for the camp in his truck, someone in Deep River would usually make a call on the one telephone line to the camp to warn them. Illegal meat was hidden in the train's huge water tank. Moonshine was also a regular "staple" at the camp.

Billy Wuorinen lived in the camp as a young boy before his parents built a house in Naselle. He remembers a neighbor who quickly buried elk meat in the ground to hide it from the warden. Not long after he buried it, a playhouse was built on the site, but he didn't have the heart to take down the children's playhouse, so the meat was left underground.

For those who grew up in Toonerville, the memories are full of love, adventure, and hard work. Billy also remembers Christmas at the camp. "We had Christmas trees with candles,

Billy Wuorinen at six years old in Toonerville. Photo from Billy Wuorinen

not many, but a few. They were only lit while opening packages. I had lots of toys, cars, trucks, toys like that. Everyone got something at Christmas."

Audrey Hollenbeck Weed remembers Christmas, too. "Christmas and waiting for "joulupukki"—(Santa Claus)—was great. Uncle Fred Hendrickson, dressed as Santa, would walk from the railroad track up the path. He'd come with a lantern or flashlight, ringing a bell as we watched from the windows waiting for him. We had to perform for Santa, a song, a dance, a poem or something, or no presents. One poem I remember that my mother taught me was, 'The Snowman's hat was crooked, and his nose out of place, and several of his whiskers had fallen from his face, but the Snowman didn't notice, for he was trying to think, of a New Year's Resolution that wouldn't melt nor shrink.'"

Sisters Mary and Betty Hill ride their Uncle Fred's speeder, a three-wheeled car built to carry people and goods on the railroad track. Fred Hendrickson used it to transport cans of milk on the tracks. He lived on the mainline between Deep River and Salmon Creek. Mary lived in the Toonerville area between 1929 and 1934. In 1965 when her Uncle Fred died, Mary acquired the speeder and stored it in her garage for more than thirty-five years. Stanley Rangila restored it for her, and she donated it to the Appelo Archives Center in January 2011, where it is on permanent display in the logging area of the museum's second floor. Photo from Mary Hill Cornell

Audrey's niece, Mary Hill Cornell, also remembers Christmas at her Grandmother Puskala's farm in Deep River. "My memories of Deep River begin when I was 4, living in Seaside, and we would all go over to Grandma's for Christmas. Everyone came. The tree had real candles on it." Just below the Puskala farm was the site of the first log dump in Deep River, just above the town of Deep River. The log dump was later moved lower down just to the west of the town site.

Doris Pearson remembered life as a young bride in Toonerville with her husband, Bertell. "When Bertell first took me up to see the camp and I saw those little 5-inch chimneys sticking every which way out of the roofs, I at first thought they were chicken houses! Most everyone kept their houses neat inside and even ordered wallpaper." A family with several little girls moved into a house across from theirs. Each night a little girl named Donna would knock at their door around suppertime and beg for cookies.

Inez Stinchcomb with Sandi and Bonnie Wirkkala. Inez did not have children of her own, but her home was always open to everyone else's children in Toonerville. Her husband, Dewey, was the train's fireman. Photo from Archives Collection

"We gave her whatever we had, and she'd put it in her little apron pocket and leave to knock on the next door in camp. One evening around dark she was called home, but had disappeared. Everyone in the camp turned out to look for her. Finally, someone thought to look in the Bowman's outhouse and there she was, sitting on that special little hole with armrests they had made for their children, sound asleep, cookies in her pocket!"

Inez Stinchcomb, wife of the train's fireman, Dewey, wrote of their arrival in Toonerville in 1932. "The house was spotlessly clean because my husband had taken the steam hose from the locomotive he set right in front of our house, and had boiled out the place before. It was sanitary to say the least! There was no paper on the walls, which had been painted before the steam hit them. There was still some paint left. We had only been there for an hour when there was a knock at the door. We were just putting the rug in place. Our first visitor came early! It was one of the loggers who lived a mile and a half below the Camp. He lived on the road. He was on a little 'speeder' contraption made with a motorcycle motor, three wheels, and an A-frame. One wheel ran on one rail and two on the other. A board about 12 inches wide and 4 feet long made the seat. They could really go, too.

"Train brakeman, Bob Sanchez, came to our door and said he was going to Deep River for groceries and wanted to know if we wanted anything. He was going house to house to ask people as was the custom everyone

practiced, since it was 12 miles to the store. Well, we didn't have one red cent to buy with, so we said, 'No, there was nothing.' 'Don't you even need a loaf of bread?' We had to tell him we didn't have the money to buy even a loaf of bread. That was easy for anyone to understand during those depression days. But I will never forget his next words. He said, 'Why the store will be glad to send you some bread and charge it for you.' C. A. Appelo was the storeowner, and he did send the bread and he did charge it! He didn't even know our name so he charged it to 'the family in the house by the engine.' That was the locomotive that sat in

Burnell Olsen's speeder, a small railroad car, was used to transport workers and goods between the town of Deep River and Olson Bros. logging camp. Agnes Paju Appelo rode the speeder regularly to the camp to take orders for goods and food. She can be seen inside the door of the speeder holding her "famous" basket. She had two large baskets she used to carry goods to the women she regularly visited in the camp. Photo from Archives Collection

C. A. and Agnes Appelo in later years. Photo from Carlton Appelo Collection

front of our house. By the way, the bell of that engine is now in Disneyland.

"Agnes Appelo, wife of C. A., came up to the camp twice a week to take orders for groceries. Sometimes she came on the train, and other times she drove up to the end of the road and walked up to the Camp visiting folks and taking orders. On the following day, the orders were sent up on the train, which went to Deep River to dump logs. Anyway, Agnes came to our house for the first time. I explained to her our situation and said we couldn't even pay for the loaf of bread sent up by Bob.

"Her words were, 'You still have to eat, don't you? If you don't get it from us, where will you get food?' That was a good question. She learned our name so it could be charged to us and up came even more than we had ordered!

Stella Kallunki Norgaard in 2011 during a visit to the Archives Center. Bertroch photo

"The nearest banks were in Skamokawa and South Bend. There was a small paycheck for those who were working, but they had no way to cash them, so what did the Appelos do? They became a bank for the folks in Toonerville. On payday, Agnes would come up from Deep River with the money to cash the checks. She carried it in her little grocery basket on her arm. After there was work for everyone in camp and times began to improve, she often had several thousand dollars with her. But not once in all the many years that she came was she molested. The people never said anything to outsiders about her banking because we all felt respect for her. She carried a small pistol with her, but it was at the bottom of her basket with candy bars and orders on top." Agnes' son, Carlton, remembers, "Mama did carry a pistol, but she always told us it wasn't loaded."

Stella Kallunki Norgaard told Carlton Appelo in 1994, "Everyone looked forward to Agnes coming. She carried goodies in her basket, covered with a towel. Besides other things, Mom always looked forward to Agnes' yeast cakes the ladies needed for baking bread. No matter the weather, Agnes would come. We enjoyed her visits besides the orders she took. In the winter months, Dad took the speeder once a month to buy groceries to last till the following month. Some winters were long and rough. In the summer we'd go to Deep River to

The Maple Grove School served children from the Toonerville and Upper Salmon Creek areas. The school is to the left. Teacher's cottage and bus garage are to the right. Photo from Archives Collection

see the movies. Pearl and Andrew Mattson ran the theater. She took tickets and he ran the projector. Those first movies were something to remember.

"There were always things planned for entertainment. There was a large hall at the end of the camp that was vacant. Everyone got together and cleaned it up so it could be used for gatherings and dances. They fixed fishponds for the children with prizes, and the ladies prepared lunch baskets, so the men would bid and [the] highest bidder got to eat with the lady whose basket they chose. Some went quite high. Then the dancing began. I started dancing when I was little with my Dad, Uncle Toivo, and George Hill, and of course, Christopher (Sparky) Yates. He was married to Gladys Salme Yates.

"There were birthdays and always parties with more pictures taken. I get a kick out of those pictures now. Some of the ladies' hats were so fancy and then the men in their long bathing suits in some photos. The parties were fun with homemade root beer in the summer and such good cakes. In the winter when we had snow, Uncle Toivo Hill would pour water on their long sidewalk in Deep River that led way down to the creek, and it would freeze overnight and we'd all rush there with our sleds. Then, we'd go indoors for a pot of chocolate with marshmallows to warm up.

"We were a close-knit family in those years. There were the Fred Gedenbergs with Thelma, Stella and the twins, Leonard and Harold. The George Hills with Betty and Mary Ann. Arvid and Jennie Olson with Vera, my best friend to this day. She is one week older than me. The Julius Lofgrens with Ella, Carl, and Harry. Also Fred Hark and Marie with Leonard and Bobby. The Walter Johnsons. The Pete Magnusons. Bertha had Faye later and they moved to Rosburg. Then Bertell and Doris Pearson came later—such a loving couple—and could he play the accordion! They brightened the whole community! Stewart and Pearl Hood came with Buddy and Lucille later on.

"We had power from the camp, and at 8:50 p.m., the lights would blink twice, and at 9:00, they would go out, so we'd scramble for our kerosene lamps to get ready for bed. We had one of the first radios, and the men would gather to listen to the boxing matches. The women had sewing bees, as they were called, and played cards on certain days.

"I'd just started high school in Naselle when our house burned to the ground. Mom had a trunk in the back room that she asked the man from the train to get out for her. He tried to spray water from the train to put the fire out, but to no avail. They couldn't budge the trunk, so Mom dashed in, and how she ever pulled it to the porch, no one could understand. Dad was working in the woods at the time. The trunk had my grandmother's (Mom's mom) items in it and some of Dad's war memorabilia, medals, hat, leggings, etc.

"In the earliest years, the Olson Camp School was located in Toonerville itself. When a road was built into the Camp in 1929, that school closed, and students attended the Maple Grove School that already served families in the upper Salmon Creek area. A garage for a school bus was built at the school so it went up the road daily to Toonerville to pick up the children. They had to walk to reach the bus stop, but they now had

The Maple Grove basketball team in 1931. Left to right: Delmar "Buddy" Nelson, Toivo Hokkanen, Archie Larson, Eugene Kilponen, Charles Owens, and Waino Kilponen.

transportation to a larger school. Over time, the parents even built a gym at Maple Grove that still stands today, cared for by Loretta and Jennifer Smalley."

Toonerville had no doctor, so when a logger was injured, the women often were the emergency medical personnel who patched them up and held them together until they could get to Deep River to get on the mail boat to be taken over to Astoria. When Fred Pentti was seriously hurt in a train accident and lost both an arm and a leg, a small building was built for him in Deep River so he could still work and make a living. Pentti's Pool Hall, or "Pentti's College," was a popular place for loggers to gather. Built by volunteers with lumber supplied by Olson Brothers Logging, Pentti's Pool Hall was a meeting place for both loggers and visitors for many years. Located where the main road meets the "mainline" road, it is now gone, but far from forgotten.

Dr. Fritz came from Cathlamet to Deep River one day a week by horseback to "make his rounds," but even after that routine began, babies came whenever they chose, so many were delivered—if the tides and

mail boats allowed—in Astoria. If the timing wasn't quite right, some babies were born in the boat on the way. Agnes Appelo's son, Carlton, a longtime historian in the area, was born on the Columbia in a boat. He often tells the story ending with, "They told my mother to throw me back in because I looked like a little frog."

When necessary, local women acted as midwives to help with deliveries. Some of the women were very good at healing the sick and taking care of injured loggers. People died of the flu, typhoid, tuberculosis, and other illnesses not so prevalent today. Women had many children, but often lost them as well. An article in the *Skamokawa Eagle* on January 31, 1895, reports, "Mr. John E. Rull of Deep River sends us the sad intelligence that his good wife, Christina Rull, died on January 12th in childbirth. Mrs. Rull was 34 years and 2 months old. She leaves a sorrowing husband and two children, aged 12 and 10 years old. Mrs. Rull was a kind, motherly woman who helped her husband faithfully to make a happy home for their little family. She will be missed very much in Deep River where she had many friends."

Life was hard, but full of adventure, too. Many of the young boys who grew up in these families went to work as loggers themselves. Some finished the eighth grade, a few finished high school, but even so, many went into logging in their early years, some following in their father's footsteps. Many were trained in their craft by their fathers: Richard Nelson's son, Howard; Joe Florek's sons, Joe Jr., John, Ken, and Gerald; Vic Wuorinen's son, Billy; Jack Linquist's son, Mark, and Mark's son, Jeff; Crint Cothren's son, Gad, and Gad's son, Dan; Lattie Metcalf's son, Keith, and Keith's son, Brandon; Jerry DeBriae and son, Jerry Jr.; George Durrah and his son, Russ, and the names go on. Penttila, Johnson, and the biggest one of all, Wirkkala.

The Maple Grove School gymnasium as it stands today on Smalley property. Bertroch photo

Pat Wirkkala, daughter of Mr. Harvey and Mrs. Jenny Wirkkala, and her brother Willard, stand on the railroad tracks near their house in Toonerville, circa 1930. This was the end of the county road. The only transportation from this point to the main camp was by rail or foot.

When Toonerville closed, the Upper Naselle Camp opened in 1939 just over the hill to the west. It was a more modern camp with a road, on a more open, flat area to build on, and even had a truck shop, Spargo's. Since many of the camp houses from Toonerville were moved down closer to Naselle, the camp's newer houses in the Upper Naselle were built for the families who would live there.

Claire Sundquist McCollum's family stayed on their farm below Toonerville. She wrote, "After the camp moved to upper Naselle in 1939, most everyone moved out, and it was so lonesome you can't believe it. There was just nothing there, but the space was still there."

Descendants of many early logging families still live in the area, especially the Finnish families. The early families lived by old country values that still underlie local traditions and culture. From Cathlamet and the Elochoman, west into Skamokawa, and over KM Hill to Grays River, Rosburg, Seal River, Deep River, and into Naselle, then on to Nemah, the train whistles can still be heard from the jake brakes on the logging trucks coming down the same hills the early loggers cleared. The same work ethics, family-first values, faith in God, and belief in the promises of our youth support and guide our daily lives today. The early family camps set the standards and gave us the history we are so proud of in this area. C. A. and Agnes Appelo's son, Burton, has said, "Let's learn what we can from the lives and values of the pioneers. There are precious few of them left!"

Loggers and their families remain the images of the area's own "American West" mythology as there were none stronger, harder working, more devoted to their families, or invested in the Willapa Hills than the loggers who initially walked into those dark forests. Some, like Joe Florek Sr., gained a reputation as a "bull of the woods" due to their strength alone. They saw in those trees the houses and barns, schools, churches, and hospitals that would be built from the lumber they provided to the new nation they now called "home." Some have gone on to serve as county commissioners: Joe Florek, Mark Linquist, and Dan Cothren, to name more recent ones, and Brian Blake, another logger, who serves as a representative in the state legislature.

Early loggers saw the future in the work they did in the beginning and still see it today. They deserve to be honored as professionals, athletes, experienced woodsmen, smart at judging a tree to see how best to fall it while staying safe for themselves and those around them. Now fewer are required to get a job done since so much is mechanized, but all the same, the older fallers are still today—as in the beginning—the best ones to teach the younger ones just starting out.

For decades, the young ones learned from the older ones who became the legends of forestry. Today's loggers are invested in stewarding timber because as Crint Cothren's grandson Dan, a third-generation faller, says, "I love seeing trees."

CHAPTER 3

Crown at Cathlamet

by Bryan Penttila

"Crown was the most highball outfit that ever turned a wheel."

Stewart Hood, 1971

Few achievements in the world of logging have stood as unchallenged as the single day logging production record set by a Crown Willamette crew working behind Cathlamet. It happened April 16, 1935, and during the most productive eight hours of logging on record, these twenty-one loggers

Crown's Willamette steel-spared skidder in 1937, two years after it set the all-time production record when its twenty-one-man crew yarded and loaded 1.6 million board feet of logs in a single eight-hour shift. Elmer Engelson Collection

yarded and loaded 480 logs. The big cuts of fir, as thick as a man is tall, filled 130 rail cars and, according to the scaler's tally, totaled a whopping 1.6 million board feet. It was immediately recognized as a world's record and made note in trade journals and newspapers across the region.

This benchmark had been orchestrated by one of the Pacific Northwest's most notable loggers, Crown's woods superintendent Charles W. Spiering, better known as "Red." The machine that did the yarding and loading that spring day was as monumental as Red's ambition—a mammoth steel-sparred skidder that was the zenith of steam-powered logging technology. Lou Rees was foreman on the big unit, and each of his loggers possessed the "highball" mentality that made Crown Willamette's camp one of the banner outfits on the lower Columbia River. It was no place for loitering, but loggers who wanted to work for a world-class operation where the timber was big and the equipment was top-notch found their way to Cathlamet.

"Cathlamet was always known as the pride of the fleet," recalled John D. Ollsen, a Crown executive in the 1960s. "It was the biggest and the best operated" logging division of the company. The operation was long recognized as a proving ground for innovative ideas in logging and forestry as well as a crucible for up-and-coming company managers. From choker setters to "conk inspectors," the pride associated with being a Crown employee was evident during the entire span of the company's six decades of active operations in Wahkiakum County. Beyond being members of an industry-leading organization, many of Crown's workers played an active role in community and civic affairs. Crown, most former employees agree with a glint of nostalgia, was like a family.

By the time Crown Willamette became the first name in Wahkiakum County logging in the 1920s, this burgeoning forest products colossus had already been dealing in pulp and paper for over a half-century. The roots of Crown's family tree were planted in 1870 when, from his modest office in San Francisco, Anthony Zellerbach began peddling paper products in the Bay Area. At the same time, and by sheer coincidence, Henry Pittock, publisher of *The Oregonian*, opened Oregon's first successful paper mill near Oregon City. A few years later, William Pierce Johnson and others set up the Willamette Falls Pulp and Paper plant in nearby West Linn. As Johnson's mill was taking shape, another San Francisco firm, the Crown Paper Company, built a plant on an adjoining site. In 1883, Pittock moved his plant, now called the Columbia River Paper Company, to a choice waterfront location at Camas, and in so doing launched Washington State's first paper mill.

With the early 1900s came a succession of mergers that would eventually result in the mammoth Crown Zellerbach Corporation (CZ). The first, in 1905, witnessed the consolidation of the Columbia River Paper Company at Camas with the Crown Paper Company of West Linn. The new company was called Crown Columbia Pulp and Paper. Nine years later, Crown Columbia merged with the Willamette Pulp and Paper Company (the word "Falls" had been removed from the company's title) to create the Crown Willamette Pulp and Paper Company. In 1928, Crown Willamette merged with the Zellerbach Corporation to create the largest distributor of paper and paper products in the United States.

In 1924, Crown Willamette launched an aggressive timber-buying campaign in the upper Elochoman River watershed. First, in December of 1923, the paper company purchased the Bradley Logging Company's railroad, camps, and timber holdings. Then, Crown completed a deal to buy the Weyerhaeuser forestland in the area, and presently bought the timber on all the adjacent land belonging to the State of Washington. Loggers went to work in the old Bradley Camp area on the east side of the Elochoman River valley, and in a short time, crews began putting up a new Headquarters Camp along Duck Creek. From there, track crews laid the Hill Line up and over the valley's west slope.

Crown Willamette's logging operation at Cathlamet was to supply the company's pulp and paper mills at Camas and West Linn with hemlock, spruce, and white fir. The Columbia River and its tributaries provided a means of transporting the logs from Cathlamet and Crown's other logging division south of Astoria, Oregon, to the processing facilities. In addition to producing its own pulp logs, company log buyers purchased rafts of hemlock and spruce from smaller logging firms all along the river. Until the 1950s when Crown entered the lumber and plywood fields, its high-grade logs were sold on the open market to sawmills and veneer plants along the Columbia and Willamette Rivers, especially the Eastern & Western Lumber Company of Portland.

Edward P. Stamm in 1933.

From the redwood country of Northern California, Crown Willamette imported a pair of forward-thinkers who would help guide the company for decades to come. Don Denman, a quiet, levelheaded executive with an exceptional mind for business, came first. Then came Ed Stamm. A big, studious man with a frightful temper, Stamm was schooled in engineering at Ohio Northern University and came to Cathlamet as a railroad engineer. He quickly rose to logging superintendent and, by 1928, was Crown's boss logger in both Wahkiakum and Clatsop Counties.

Old-time loggers were a bit reluctant to take orders from a college-trained engineer, but Stamm quickly won their confidence. He insisted, much to the loggers' satisfaction, that the company's camps be clean and comfortable, something still considered a luxury in the 1920s. He imported the biggest and most efficient logging equipment, including the giant Ledgerwood and Willamette steel-sparred skidders. Crown's wunderkind impressed his superiors by innovating the way the woods worked. He

made major strides toward better utilizing the resource by having crews re-log cut-over areas to gather logs bypassed during the first go-round. Later Stamm helped devise "pre-logging," a more productive alternative to "re-logging" that involved removing the smaller stems from a stand before the "prime cut." Perhaps his greatest innovation at Cathlamet—and one adopted industry-wide—was introducing and adapting the bulldozer to woods work.

Crown Willamette's camps in Cathlamet were known for highball logging, and the equipment was geared to match. No. 12, a Baldwin 2-6-6-2 saddle tank, and her sister engine, No. 16, were the zenith of logging locomotive technology in the 1920s. Crown used No. 12 at Cathlamet until the end of the line in 1959. Photo courtesy of Deetta Funderburg

Stamm took a keen interest in the development of young professionals, and during his tenure at Cathlamet, a number of future CZ executives got their start. Harold Miller was one. Like Denman and Stamm, Miller worked in the redwoods before being hired as a surveyor at Cathlamet in 1927. He eventually became the company's chief logging engineer. Another was Howard "Humpy" Peterson, an ambitious Elochoman local, who started setting chokers in 1933. While working in the rigging one day, he

Crown Willamette's Camp 2. In 1936, Warner Brothers used the camp as the backdrop for its Technicolor melodrama *God's Country and the Woman*. Photo courtesy of Deetta Funderburg

wrestled Francis Edmonson to the ground and while playfully smearing mud over his prematurely bald co-worker's dome declared, "One day, Francis, I'm gonna be your boss." True to his word, Peterson advanced through the ranks and retired as a CZ vice-president.

Despite the crippling effects of the Great Depression, the 1930s were a busy time at the Cathlamet operation. Its several dozen miles of steel were kept polished by the wheels of log cars pulled by the big Baldwin mallet locomotives as they brought nearly four hundred thousand board feet of logs out of the woods daily. The last train rolled through Bradley Camp in 1928, and by 1932, logging on the Hill Line was nearly complete. Even when log prices sunk to abysmal depths, Crown kept working. The company pushed its rails well ahead of the loggers, thereby allowing the crews access to stands of different timber types that could be brought out to meet market demands at any given time. Whether they were working in pulp or peelers, logging foremen Eric Johnson, William Mattson, and Lou Rees saw to it that the cat logging, cold-decking, and skidder sides were all humming a high-ball tune.

As the timberline receded up the valley, a new camp was built at the main forks of the Elochoman River about seven miles above Headquarters Camp. It was called Camp 2. The most notable landmark at the facility was the giant machine shop and locomotive barn where the vast array of rolling stock and equipment was housed and maintained. The loggers with families remained at Headquarters Camp—thereafter known as Family Camp—while the unmarried employees moved to the new site and settled into one of several dozen bunkhouses, each equipped to house four men. Near the rows of tidy bunkhouses sat the sprawling cookhouse where as many as 250 men were fed daily. It took a team of twenty people, including cooks, bakers, meat cutters, "flunkies" (the loggers' term for waitstaff), and dishwashers, just to accommodate the hungry workers.

In 1936, Camp 2 became the backdrop for a Warner Brothers movie called *God's Country and the Woman*. William Keighley directed the Technicolor drama based on a novel by James Oliver Curwood that starred George Brent—a then well-known leading man—as well as Alan Hale and Beverly Roberts, a last-minute replacement for Bette Davis, who had been slated as the female lead. A special part was even written into the script for the movie's technical advisor, "Red" Spiering. When *God's Country and the Woman* was released in 1937, critics panned the film though, not surprisingly, it was a sensation in Cathlamet's Elco Theater.

Along with all this activity up in the woods, down on Cathlamet Slough, the din of bandsaws were making a terrific racket. In 1929, Crown Willamette built a sawmill hard by the railroad log dump to break down oversized pulp logs into "cants" that could be handled by the paper mill's chippers. A few years later, the mill was retooled to cut high-grade lumber, mostly spruce for use in airplane construction. In 1933, International Wood Products, a subsidiary of the Kraft Cheese Corporation, put up a factory adjacent to Crown's mill to make cheese boxes. Lumber for the box factory came from the Crown mill. Cathlamet remained a key producer of Kraft cheese boxes until 1941 when the plant was idled, followed three years later by the closure of Crown's sawmill.

The 1930s also witnessed rumbling of a different nature when Crown's loggers began to organize. By 1936, the Sawmill and Timber Workers' Union Local 2592 represented Crown's employees at Cathlamet. The organization's efforts went smoothly compared to those at Crown's Lewis and Clark camp a few miles east of Seaside, Oregon, where two loggers were gunned down in a riot between rival unions in May of 1936. On one occasion, recalled Crown sawmill employee Herb Faubion, an American Federation of Labor (AFL) organizer strode into Cathlamet to stir up support for his association. He promptly turned around, however, a little bloody but a lot wiser to the fact that the workers of Cathlamet had already pledged their support to a union. In 1937, the Sawmill and Timber Workers' Union became affiliated with

Harold Olsen at the helm of Crown's towboat *Constance J*. Olsen went to work for Bradley Logging Company in 1903 when he was twelve years old and carried over with Crown Willamette when it took over the Bradley operation in 1924. He became the skipper of the *Constance J* in 1941 and retired from CZ a dozen years later. Photos courtesy of Robert Olsen

the International Woodworkers of America (IWA), part of the Congress of Industrial Organizations (CIO). Soon every logger in Wahkiakum County was a member of an IWA local.

Keeping a close eye on the developments at Cathlamet from his new office in Portland was Ed Stamm, the newly appointed logging manager for CZ's Northwest Operations. His progressive notions on forest management made Stamm a leading voice in the forest industry, and for his efforts, CZ named its Columbia County Tree Farm after him following his retirement in 1959. Stamm's was not the only noteworthy departure from Cathlamet during this period. In 1937, after eleven years as logging superintendent, the incomparable "Red" Spiering left, too, to take a job outside Sweet Home, Oregon. He eventually left the logging trade and became a successful contractor around Eureka, California.

New talent was arriving at Cathlamet as well. In 1939, a twenty-four-year-old named Ray Knopski hired on as a chaser to work under the Willamette skidder that had set the world's record for log production four years earlier. Ray fit right in at Crown. He was born in a logging camp on the Washougal River and had followed the timber trade since his earliest days. Except for a stint in the Navy during the Second World War, Ray dedicated his efforts to the company and worked his way up to hook tender and, eventually, rig-up boss. In 1947, he married Inez Quigley, the daughter of an old and influential Cathlamet family, and together they made their home in Family Camp. After a prolonged labor strike in 1954, CZ officials appointed Ray logging foreman.

By the time Knopski became bull of the woods, the end of CZ's railroad operation was in sight. The company had quit building new rail line and was trucking their logs from the cutting areas to the railhead at Transfer Stations 25 and 33. In 1955, after over three decades without a fatal accident on its rail line, veteran brakeman Albert Wend died in a freak derailment a couple miles above Camp 2. His brother-in-law, Walt Witham, was at the controls of the locomotive that day. Three years later, on December 18, 1958, Engineer Witham piloted the last trainload of logs down the twenty-six-mile long mainline through the Elochoman Valley. With its railroading days complete, CZ parked its oversized and outmoded steam equipment on a rail siding to await the scrappers' torch. One locomotive, Willamette Shay No. 9, was preserved as a point of historical interest and can be seen in Cathlamet today.

Several years before the last train whistle echoed through the Elochoman Valley, CZ was gearing up to move into its next big block of timber, the Brookfield Unit. There, nearly thirty square miles of mostly old-growth hemlock and fir—over a half-billion board feet of it—stood ready for harvest. In the mid-1950s, construction crews punched a series of access roads into the area. Atop KM Mountain, a truck shop and maintenance facility went up, and down along the Columbia River, the historic salmon canning town of Brookfield was transformed into a log dump and rafting facility. Soon, Brookfield was buzzing. Jerry Bain began setting chokers near there under a wooden spar tree after he got out of Wahkiakum High School in 1959. A few years later, Bain's side got the first modern steel-towered yarder in CZ's rapid shift to the machines.

In 1958, CZ transformed the historic salmon canning town of Brookfield into a modern log dump and rafting facility. The currents of the Columbia and rogue wakes from passing ships gave the boom men at Brookfield occasional headaches, but CZ used the site for nearly two decades. Photo courtesy of Philip Penttila

As work progressed at Brookfield, CZ began opening up its westernmost holdings in the Naselle area. In 1961, CZ revamped Deep River Logging Company's railroad log dump in the town of Deep River. Nearby, truck scales were installed, and just up the road, a cavernous truck maintenance shop was built. From Deep River a first-rate mainline logging road was built into the company's holdings on the upper Naselle River. Giant off-road log trucks were soon rumbling down the hill with tandem trailers in tow. The idea for tandem trailers, it should be noted, came from CZ chiefs Howard Peterson and Clarence Richen.

In the summer of 1962, G. W. "Wally" Nutter replaced Dean Prater as

Inset left: In the early 1960s, CZ began replacing its wooden spar trees with steel-towered yarders. Photo courtesy of Jerry DeBriae

Left: Stanley Rangila at the controls of a Washington TL-6 loader on Altoona Ridge. Notice in the background Grays Bay and the mouth of Deep River.

With its extensive network of forest roads, CZ employed giant off-road trucks with tandem trailers to haul its logs to the dumps at both Brookfield and Deep River. Here, a driver secures a binder on a load at the Salme Ranch along the upper Naselle River before the trip to Deep River.

division manager. Nutter was reared in a Grays Harbor logging family and worked his way up the high-lead hierarchy for Crown Willamette in the hills behind Astoria. He was new to his post at Cathlamet when the infamous Columbus Day Storm tore through the region on October 12, 1962. It decimated well over one hundred million board feet of CZ's timber and left contract logger George Oller with a jumble of big sticks strewn over his equipment. The storm launched CZ into the log export market and resulted in a significant uptick in logging activity.

To make sure every log reached its proper destination, in 1963 the company built a sorting yard just west of Cathlamet. Logs came into the yard both by water—from the dumps at Brookfield and Deep River—as well as by truck. The big sticks were handled by equipment of an equal scale, including a ninety-foot electric stiff-leg crane and giant LeTourneau and Wagner log stackers, capable of picking entire truckloads in a single lift. Once spread in the yard, the logs were scaled and separated into any of

forty different sorts, depending on length, grade, species, and customer. Don Klecker, and later Jack Lindquist, had charge of the operation, which some days handled five hundred thousand board feet of logs.

CZ increased its harvest volume throughout the 1960s, and by the end of the decade was even logging at night. On several of its yarders, the company had replaced the butt rigging and chokers with one-hundred-inch grapple carriages. It was largely a hands-free affair in the brush, requiring only a radio-wielding spotter for the grapples and a hook tender to move the blocks. CZ officials were not tardy in sensing that such an operation could be conducted at night. One crew worked the machine in the daylight, and as they prepared to go home, the night

Employees at CZ's sorting yard in Cathlamet place steel bands around a load of logs to prevent "sinkers" from dropping out of the bundle while it is in the water. In the background, the ninety-foot electric crane (previously used in the construction of the bay bridges near San Francisco) that was used to hoist the logs from the Elochoman Slough. Photo courtesy of Deetta Funderburg

Above: In 1961, CZ reconfigured Deep River Logging Company's railroad dump at Deep River to serve log trucks. The firm continued to use the facility until the early 1980s. Photo courtesy of Jerry DeBriae

Left: A look upriver at the log dump at Deep River in 1979. The two little boom boats were used to push bundles of logs into the raft at left.

shift arrived. With banks of mercury vapor lights mounted to the yarder and another set hoisted above a rubber tire skidder, the night loggers went to work. For the next eight hours, the night crew snagged logs with the giant grapples and piled them high at the foot of the yarder. For over a decade, CZ maintained its night logging program.

During that time, no family contributed more to CZ's night and day shifts than the

Cothrens. The clan originated in North Carolina and worked in the Ryderwood area for Long-Bell Lumber Company until the mid-1950s when they began a great migration over the hill to Cathlamet. Among them were a passel of brothers and cousins and sons that included Red, Crint, Kenny, Carl, Drew, Royal, Loman, Gad, and G. L. Many of their sons, too, took jobs with the company, and over the years, there were no fewer than twenty Cothrens logging for CZ. Many of them worked in the rigging, but Crint joined the cutting crew as did his son Gad. In the 1970s, Gad's son Dan also became a cutter for CZ and by 1986 had become a bull buck.

By the early 1980s, CZ was an ailing corporation. Stockholders had endured years of lackluster dividends that some analysts claimed was due to the company's top-heavy, outmoded ways. When the bottom fell out of the forest products markets in 1980, CZ began to restructure. At Cathlamet and the other woods divisions, it began a transition from paying its employees an hourly wage to a pay-for-production setup. Timber cutters had been on a similar wage-incentive program already. Most loggers saw a small uptick in their earnings, and the company realized a boost in production. That, however, could not cure the multinational corporation's ills.

One of the world's richest men saw a profit to be made from the staggering giant. His name was Sir James Goldsmith, an Anglo-French financier who had made a name for himself as a corporate raider. Late in 1984, he began aggressively buying CZ stock and, by the following summer, owned 52 percent of the shares. Once in control, Goldsmith quickly disposed of the company's paper mills and focused on Crown's most lucrative jewel, its 1.6 million acres of forestland. Soon it was announced that CZ would phase out its logging operations as part of Goldsmith's new restructuring strategy.

The dismantling of CZ and the disappearance of its payroll left local communities to ponder an uncertain future. Weyerhaeuser's closure of its Grays River operation in 1983 and CZ's shutdown a few years later were weighty events but part of a larger trend. In Wahkiakum County, employment in the logging sector had decreased by over half between 1969 and 1983. Most of the job losses came after 1977 as the last major stands of old growth fell in the Willapa Hills. Harvesting big trees required big crews, and as the nature of the timber changed, so too did the workforce requirements.

Towns like Cathlamet and Naselle survived, of course. Although the area lost its two primary employers in a jarring two-year period, the resilience of the population met the challenge. Some jolted CZ employees found jobs elsewhere, others took an early retirement or found work with the handful of local gyppo loggers. Wherever they landed and whatever else they did, most former employees are quick to note the Crown Zellerbach was the best damn outfit they ever worked for.

Ray Knopski, CZ logging foreman from 1954 until 1979. Photo courtesy of *The Wahkiakum County Eagle*

CHAPTER 4

Crown Family Camp on the Elochoman River

by Karen Bertroch

> *I had a great childhood. I've seen a lot of changes in my years. My good old days really were good old days. There is so much to tell about living in a logging camp. I guess I was a lucky one.*
>
> Rowena Ehrlund

In response to the increasing lumber boom of the early twentieth century, one of the largest camps in Wahkiakum County, Crown Willamette's Family Camp, was opened early in 1925 on Duck Creek up the Elochoman Valley north of Cathlamet.

> *During 1925–26 the company decided to expand its camp to include housing for family men. In 1927, there were 23 homes at the site. In 1929 another 22 family residences were added. The houses had three rooms in 1927, then three and six rooms in 1929. The newer ones were equipped with indoor plumbing and electricity, unusual in any logging camp of the day. The shop, office, cookhouse and bunkhouses from Bradley Camp on Beaver Creek were all gradually moved to Camp 2. By 1936, only the homes of the married loggers remained at the Upper Family Camp, as it was later named. Accommodations for up to 250 unmarried loggers were available in the late 1930s in 70–80 buildings below Family Camp.*
>
> *Family Camp's houses ran along the north side of the road and railroad spur that ran up Duck Creek, a short tributary to the Elochoman. Built on a gently sloping hillside, the 1927 houses were arranged in two tiers overlooking the rail sidings and auto garages which occupied the flat valley floor below. Family houses were uniformly cut, long and narrow structures, 14x38 feet. Typically, the family houses were three-room units with a kitchen, living room and bedroom. The few six-room houses were made by putting two 3-room units back to back and covering them with a single roof.*
>
> *Across from the shop stood the office, guest house, first aid room, and commissary. Behind this cluster were some 60 dwellings arranged in long rows. The large cookhouse was in the*

Next page: Crown's Camp 2. Photo courtesy of Deetta Funderburg

Crown Zellerbach Camp 2. Photo from Archives Collection

center of the bunkhouses, a low rambling, one story building with two long wings lined with windows. The bunkhouses were simple 4-man cabins with outside walls measuring 12x16 feet, constructed of tongue and groove siding with one small window in the center of each of three sides. The 4th side held the windowless door which opened to an outside faucet and 8-foot breezeway paired with another cabin. All the units of the camp were connected.

K. Erickson, Ph.D., University of California, Berkeley, 1965

As years passed and log transport modernized using off-road, then highway, logging trucks, companies were doing well, and some realized they would have a good return if they invested in their employees' welfare with housing and benefits. Crown Willamette, now Crown Zellerbach, was a company that chose to continue to support its workers and their families as long as possible.

But in 1958–59, Crown Zellerbach shifted its logging operations to Brookfield, west of Cathlamet on the Columbia River, and vacated Camp 2. When they closed, homes were offered for sale to employees willing to move them to new locations, and today an assortment of dwellings, outbuildings, storage sheds, and small barns scattered over Skamokawa, the Elochoman Valley, and Cathlamet can trace their origins to Camp 2. On December 18, 1958, when the last load of logs hauled by rail moved down the Elochoman Valley to the Cathlamet log dump, an era ended.

Brookfield seen from above in 1978. Dock can still be seen at top of picture. Photo courtesy of Rod Smith

Logger Wayne Holloway from Cathlamet remembers logging for Crown up the Elochoman. "After getting out of the service in '57, I got a job with Crown. I took the speeder up at 4:30 a.m. It picked up guys along the way. We went to Family Camp and Camp 2 and picked up more. I never lived at the Camp. They lived pretty good up there. There was also a Camp 3 in the earlier days. The winters in those days were pretty cold, long and tough. When it started to snow, the loggers went down to the valley for the winter."

Wayne hauled out the last load of logs for Crown in his self-loading truck. He recalled bosses he'd worked for. Some were loud, yelling and screaming. Some were so big and strong they could handle a two-man saw alone. Joe Florek was one of them who could handle one by himself. Leslie Wirkkala could, too.

He remembers classmates who didn't fare so well. Some families lost more than one son in the woods. Before workers had to be eighteen to be hired, there were boys who left the eighth grade and went to work with their family. Wayne recognizes the changes. Oxen were replaced by steam donkeys. Splashdams gave way to trains, and roads replaced the trains and tracks. Off-road trucks made way for highway trucks. And walkie talkies gave way to cell phones.

Kayrene Gilbertsen from Puget Island reports, "I remember a story from Elizabeth Foster Doumit, who grew up in the Elochoman Valley. She said some of the men boarded at their farm during the winter and

Crown Camp Cookhouse where up to two hundred men were served at every meal. Photo courtesy of Deetta Funderburg

would do a variety of work for her father. She was little but helped her mom with cooking for all the men. That alone was quite a chore."

Debby Heywood Halliburton was a "city kid" from Portland who visited her grandmother, Della Quigley, during the summers at "my Grandmother's little house on the highway. Almost all of my uncles and my grandfather, Leonard

Quigley, were fishermen and loggers—a combination that I don't think is very unusual in this area. Grandpa Len added trapping to that mix as well. He was the father of 11 and had to pursue just about every source of income he could! His wife, my grandmother Della Souvenir Quigley, was born in the Elochoman Valley in 1885 and spent almost all of her 98 years in Cathlamet. She was honored at Crown Zellerbach dinners as the wife/mother/grandmother of so many Crown employees.

"The logging industry involved so many people in those years, from timber cruisers to the actual working crews, to the men who worked at the log dump sorting and jumping around all those dangerous floating trees. I never lived in Cathlamet, but I was lucky enough to spend summer vacations in Grandma's little house on the highway, close enough to walk to the slough and watch the men run the donkey and dump the logs into the water. We loved the sound of that whistle!

"[I] [r]emember the women in logging, too. My aunt and uncle, Carol and Frank Quigley lived in Skamokawa and logged their own property. He also worked for Crown at various times. When they logged for themselves, my aunt would drive the log truck and pass my Grandma's house on the highway with a full load, always blowing her air horn as she passed by. Uncle Frank bought an old logging camp house and had it moved into his backyard in Cathlamet.

Leonard Quigley on the right is with an unidentified man after they've caught a sturgeon in the Columbia River. Photo is from Leonard's granddaughter, Debby Halliburton.

"My aunt and uncle, James and Delores Cochran, lived in the Crown Family Camp in the 1950s when I was very young. We would visit them there. All the kids loved to play at the end of the street in a little playground. One time I decided to go back to the house on my own. I walked down that long row of houses, climbed the stairs and walked in the front door. No one was home—where were my mom and dad? Aunt Dean and Uncle Curly? They were sitting in the kitchen playing cards when I left. I wandered through the house and saw a picture on the wall. Who were those people? It suddenly dawned on me I was in the wrong house! I ran out and continued down the road till I saw our car in front of another house. I got it right that time! The woods were so much a part of my life in Cathlamet and so interesting to a 'city kid' like me."

Carol Cothren in 2011. Photo by Bertroch

Carol Cothren has some Family Camp stories. "Gad, my husband, was a 'tarheel' from North Carolina and logged at Ryderwood where we met. We married in 1952 when I was 17, and our first child, Dan, was born in Ryderwood. When Long-Bell closed, Gad went to work for Crown Zellerbach. We moved to Family Camp on the Elochoman in 1955 and lived there till 1960 when we moved into our own home on Puget Island.

"Family Camp was on Duck Creek several miles up the Elochoman Road. It's called Crown Camp Road now. I'm sure there were camps long before ours, but this one was large and more modern, although in reality it was a two-bedroom house heated with what we called 'tin heaters' that burned wood. We had an old-fashioned wringer washer and a wood cookstove. Sometimes in the winter, there would be too much snow to do any logging, so no work, but we were happy to have a home to live in. We did pay rent, about $10 a month. Crown provided firewood and hauled out garbage for free.

Left: Off-road trucks carried huge loads of logs. This double load illustrates the loads one truck pulled. They rarely went faster then thirty-five miles an hour and stayed off the highway as much as possible. Howard Nelson of Grays River, logger and longtime truck driver, remembers hauling double loads from Nemah through Salmon Creek to the mainline road, down to Deep River and onto the off-load area where a crane lifted logs off the truck, then dropped them into the river. Drivers then turned around and went back for another load. Photo courtesy of Jerry DeBriae

The Deep River log dump area in winter snow. Pentti's pool hall can be seen in the background. Photo from Archives Collection

"Those were good years. Mrs. Wansor had a private store and offered credit to the families. She took the kids to Sunday School in town. Crown provided insurance. Cathlamet had a movie theater, and the place where the Eagle newspaper office is now was a restaurant. There were bars, of course, too.

"It was a good life. Crown was booming. Fishing was good, too, especially when the salmon were running in Duck Creek. We were all like one family at the Camp. In tough times, we all shared. No one had any more than anyone else. No one felt as if they were 'less than' anyone else either. We all shared what we had and made sure everyone had something to eat in the winters when money was tight.

"Gad did everything—choker setter, faller, chain filer. Once Gad twisted his knee. State L&I paid, and that helped, but it was slow. He started out as a choker setter. He and his brother were really fast. Toward the end of Gad's working years, he drove logging trucks to Warrenton and Longview. He was a good worker and a good provider, a good family man who loved his grandkids. Work and jobs were plentiful during those years.

"When we lived at Family Camp, the kids played games outside all the time. Kids were pretty healthy, although there was hepatitis, and one boy got polio, but he was the only one. Dr. Avalon took good care of loggers who got hurt. In the mornings, the women went house to house to visit in their housecoats, even when it was cold. Kids rode the school bus to school. There were around 16 houses in the area we lived in and 10 more in the lower camp where the managers and woods bosses lived—100 total people at that time.

"In the summers, they went to work really early so they could avoid the heat and humidity in the afternoons because it was so dangerous. Those conditions added to some sort of fuel, like a spark from cables rubbing on bark, could cause an explosion, so when the weather was hot and humid, they ended their workdays in the early afternoons. Those were the 'hoot owl' hours. When the men took off so early in the mornings, the owls were still hooting! We took old growth for granted because we lived in the middle of it."

Rowena Knopski Ehrlund commented, "Living in Family Camp for me and my sisters was a very special time in our lives. We were friends with them all—more than friends. We were family. Mom never knew who we would invite to lunch. We just brought them home with us. Our homes were row houses with wooden sidewalks. Some were three rooms plus a bathroom, some had two bedrooms, and some had three bedrooms. Most had oil or wood heat.

"The Camp provided a playground up by the wood yard where the train brought wood logs to make firewood. The kids were told to stay off the logs, but being kids, we were always there playing on the logs and sawdust and railroad tracks.

"The Camp provided garages for our parents' cars. We had a 1950 Desoto, Fluid Drive, 2-door, blue with grey wool seats. Before that, they had a car with curtains since they had a new baby, me.

"The big flat car speeder used to take the men to work and bring them home on the train tracks. When the men were left off, the kids would go and meet their dads and see what they saved for them in their lunch pails. Hunting season, the men took the speeder into the hills. Elk hunting was kind of a new thing. When they came back, the speeder was full of elk, only cows and calves. The people of the camp needed the meat.

"Summer was fun, no TV. We had a party telephone line that you cranked. We played outside all the time, morning to dinner time. Lots of friends to play with. We played in the nearby creek and we peeled bark along with the parents, for the money.

"In hot weather, the loggers were on 'hoot owl' schedules. I hated that. It was hot, and all of us kids wanted to play, but our dads had to go to bed early because they got up so early to go to work. The mothers tried to keep us quiet. Mothers didn't work outside the home. Keeping us kids in line was a big enough job.

The Cathlamet sorting yard boom that lifted logs into the Elochoman Slough where they were made into rafts and rafted out to the Columbia and out to a sawmill. Jack Linquist was the man in charge of the boom and sorting yard. His son, Mark, in partnership with Jerry DeBriae, later ran the yard for the chipping company they owned together. Photo courtesy of Jerry DeBriae

Early years. The boom was simply made of trees. Photo courtesy of Jerry DeBriae

"In the winters, our dads didn't work if the snow got too deep, so they played pinochle at night at the neighbors. When things were tough, Mom might get up in the morning, and in the porch light might be a fresh package of venison. No one knew who brought it; no one asked. Whoever needed the meat, well, we all shared; we were all one family.

"As a young person, I learned to hunt and fish with my dad. I rode on the speeder, the log train, and sometimes ate with the bachelors in the main work camp where the train shop and offices were.

"I learned so much about the people in the log camp and the Elochoman Valley. I learned what true friends are. These people are still very important to who I am today. So many are gone now, but I can still see their faces. I remember

The Crown boom area at the sorting yard. Logs are sorted into rafts. Photo courtesy of Jerry DeBriae

so much, like taking my wagon and going down the road to my cousin's gramps to get fresh milk. I had a great childhood.

"I've seen a lot of changes in my years. My good old days really were good old days. There is so much to tell about living in a logging camp. I guess I was a lucky one.

(From 'To Muumu With Love' by Rowena Ehrlund, 1986)
Money comes hard and conditions are rough.
We have seen children become fatherless.
Tears have been shed for those we have lost in the woods,
But they would do it over again even though it hurt so many of us.
The woods are their life no matter what,
God protect the rest of them, please, for us."

Janet Souvenir Bryan grew up at the Camp just a few houses from Rowena's. She remembers, "We didn't have a lot but we were happy. Kids were outside all the time. I liked to play in the creek with pollywogs, frogs, and crawdads, and I made mud pies. I was dirty a lot but it didn't bother me! We swam when water was high enough in the Florek's ditch. I remember Joe Florek Sr., such a nice man. They had a gentle, kind family. And our bus driver, Louis Jaspers. He and I had the same birthday. He always brought me jelly beans on my birthday. He never forgot.

"We didn't sit around and watch TV. We were always outside using our imaginations. We had forts and playhouses in the trees around the Camp. A cluster of trees would become a room. I remember Paul Cothren; he was a hoot. Both he and my sister, Leota.

"The main thing I learned there was how to be family. Kids could go to anyone's house to get what we needed. We were just one big family. I think Cathlamet is still like that. If someone gets hurt or needs help, we all do what we can to help.

"Everyone was a good cook. We always had people at our house. Dinner was family time. We had real butter and homemade ice cream. Food was a big part of my childhood. On holidays, we always went to my grandparents' house. Grandpa Souvenir insisted that after the women cooked, the men would do the dishes, and they did!

"My mother, Marie, cooked everything from scratch. We ended up eating lots of fish because they were so plentiful. There were lots of gardens, and everyone shared. We also had pigs and chickens at my Grandad Gachstatter's nearby farm. I learned to cook from my mother. Here is one of my favorites she used to make, German Hotcakes: Combine 1/2 cup flour, 3 eggs slightly beaten, 1/2 cup milk, 1/4 teaspoon salt. Bake in 2 tablespoons melted butter in an iron skillet for 15 minutes at 450 degrees.

"We were happy. We learned as kids that material things aren't that important. We learned to appreciate nature and learned it was important to spend time with family."

Dan Cothren, local faller and county commissioner, graduated from high school in 1971 when he was eighteen and went right to work for Weyerhaeuser. A short time later, he went to work for Crown. "My dad, Gad, had me out cutting and working in the woods when I was real young. My dad worked for Long-Bell starting when he was 21. It was hard to get timber workers at that time. Grandpa, Crint Cothren, recruited workers in the '50s.

"We lived at the Crown Family Camp. I remember my dad coming home from work when I was little. Mom made Spam sandwiches for him, and I loved them. He'd leave a little of the sandwich for me. I loved to meet him when he came home so I could get the sandwich. Times were tough in the winters with no work and no unemployment either. But the union got us commodities as I remember. Cheese, flour, beans. Sure helped.

"There was a store at the camp run by Mrs. Wansor. She spanked me once. The store was at the Upper Camp. Upper Camp kids sent me up her apple tree to shake apples down for the high school kids. She caught me and gave me a spanking I never forgot!

"I loved it when Dad took me up to the Upper Camp where the cookhouse was. They had a famous cook up there, Henry Johnson. When Cathlamet used to have a smorgasbord to raise money for the schools, Henry was head chef. People came from all over to eat his food. When Dad took me with him to the Upper Camp, he would leave me at the cookhouse with Henry, who'd give me a piece of his apple pie and ice cream. What a treat!

"Everyone had credit from the store. We all went from payday to payday. New ones learned, and we just took care of each other. I was outside all the time, usually in the creek playing. I'd get new shoes and immediately go get them wet. Sure made Mom mad!

"I remember my Grandpa would take pieces of kindling to stick to his cork boots. He walked down to the Camp because the speeder would stop there. When wood stoves in the Camp houses got too hot, they'd jump right across the floor! Honest!

The Cathlamet Tree Farm sign with the Crown logo and "Trees Forever."

"I'm proud of being a logger. We're professionals, highly skilled. We felled it; we bucked it. I'm proud of that."

Crown's train, the Six Spot, is on permanent display at the Wahkiakum County Historical Society's museum in Cathlamet. Saved years ago by members of the community, it may be caged, but it has run for miles and miles and deserves a good rest where visitors can learn of the logging history of the area. Photo by Bertroch

CHAPTER 5

Weyerhaeuser's Grays River Operation, 1955–1983

by Bryan Penttila

Getting called back to camp during the workday never meant good news for Weyerhaeuser loggers working at the company's Grays River branch. It usually signaled that someone had been hurt or, worse yet, killed. So when instructions were broadcast over the company radio to return to camp just before noon on February 25, 1983, employees' hearts sank—they all knew something bad had happened. Once the men were gathered in the truck shop, the only building large enough to hold all 190 of them, a Weyerhaeuser executive thanked them for their efforts and then announced that the camp would be closing. A sense of shock swept the employees, for there had been no talk of a shutdown. The company, explained the official, faced an over-supply of wood in a spiraling market, so the decision had been made to idle the operation. The men were given the rest of the day off with pay. That afternoon and well into the night, the nearby Grays River Tavern roared.

In its near three decades of operation, Weyerhaeuser's Grays River camp had become something of a local institution. It was one of the Naselle-Grays River area's primary employers, where college-bound youth spent summers earning tuition dollars and career-minded loggers found stable and high-paying work. The company offered scholarships to area high school students and contributed generously to community activities. So ingrained in the local community was the timber giant that some locals still refer to the upper Grays River basin simply as "Weyerhaeuser."

Weyerhaeuser had been the biggest name in the woods of western Washington since January of 1900. It was then that St. Paul's Frederick Weyerhaeuser led a syndicate of midwestern lumbermen in the purchase of some nine hundred thousand acres of old growth from the Northern Pacific Railroad Company. The agreed-upon price was six dollars per acre, which, not surprisingly, turned out to be a fantastic bargain. The newly formed Weyerhaeuser Timber Company kept buying forestland and by 1903 controlled 1.3 million acres in Washington State. The directors of the concern were in no rush to enter the logging and sawmilling

Next page: The twelve-foot two-inch yellow fir felled along Grays River's East Fork in September of 1955 by Vern Larson and Burl Broyhill. Kneeling, left to right: Julius Conrad, George Fauver, and Charles Keiski. Standing: James Blankenship, Willard Salme, Ray Ervest, John Bighill, and Otto Keiski. On top of log: John Haglund and Walt Paavola. Photo courtesy of Robert Paavola

A Weyerhaeuser "cat logging" operation of the 1950s. Photo courtesy of Keith Metcalf

business, opting instead to operate as a timber-holding organization that sold stumpage to smaller, independent loggers and sawmills.

That notion went up in smoke—along with tens of thousands of acres of the company's forestland—in September of 1902 when the Yacolt Burn swept across the headwaters of the Lewis River. To salvage the charred trunks while they still possessed some value, Weyerhaeuser dug deep into its pockets and hastily built a railroad and camps. Soon they were logging in a big way. For the next twenty years, the company salvaged the burnt sticks and cut a fair amount of green timber standing nearby. In the mid-1920s, as logging around Yacolt began to wind down, Weyerhaeuser drew up plans for an even larger operation at Longview, a sawmill city recently built by the Long-Bell Lumber Company of Kansas City. There, in 1929, Weyerhaeuser opened one of the world's largest sawmill facilities and, four years later, a sprawling pulp mill.

During this time, Weyerhaeuser expanded into other parts of southwest Washington as well. In 1931, the company organized a subsidiary called the Willapa Harbor Lumber Mills and purchased two sawmills in Raymond and another in South Bend. The Great Depression had hit the communities on Willapa Harbor particularly hard, and the arrival of Weyerhaeuser was a welcome relief. Soon the mills were up and running, and, in the hills north and east of Raymond, an army of company loggers worked to keep the log booms full. Across the Willapa Hills, near Pe Ell, Weyerhaeuser geared up for a mighty logging show with the opening of Camp McDonald in 1939. This camp became the headquarters for an expansive logging railroad that threaded through the hills at the headwaters of the Chehalis River. By the time of the Second World War, Weyerhaeuser had major operations on three sides of the Willapa Hills.

Not long after the war, the Wirkkala Brothers Logging Company began making the first inroads in the upper Grays River valley. An earlier generation of loggers had removed small parcels of timber from the headwaters of the river using splash dams, but by the late 1940s, the roadless watershed contained the last big stand of big trees in the area. The brothers—Albert, Ben, and Paul—had been making stumps on state land, first in section 36 (T11N R7W) and by 1951 in section 16 (T11N R6W). Getting into the stands required pioneering a truck road up the river valley, high above the vertical Grays River gorge, and a pair of wooden bridges to span the stream. Below the gorge, in a field on the old Cook family farm, the Wirkkalas built a logging camp replete with a truck and blacksmith shop, filing shed, and office. Some twelve miles away at Deep River, they constructed a log dump near Svenson's Corner. With the state timber logged and access roads in place, the brothers soon had a contract to cut Weyerhaeuser wood.

For a number of years, the Wirkkala Brothers got along handsomely in Grays River, turning out great quantities of large, clear peeler-grade fir for Weyerhaeuser's new plywood mill in Longview. Then, in early 1955, Weyerhaeuser made the Wirkkalas an offer they could not refuse. For a reported three quarters of a million dollars, the Wirkkala Brothers sold to Weyerhaeuser their camp, equipment, and dumping facility. The Grays River works became part of the company's Longview Division where Logging Manager Clyde Corman had charge.

Corman, Nelson Jeffers, and other key personnel immediately set to work developing a cutting schedule and logging plan. The new operation encompassed some seventy thousand acres, though only about a third of the landmass contained "mature timber." Preliminary cruise estimates put the volume of old growth at nearly 1.2 billion board feet—enough wood to keep the company's loggers busy for about thirty years, one report indicated. The cruisers also noted that in some areas the "decadent" timber was quickly losing its volume and value owing to deterioration. Something would have to be done about that, and quick. The logging plan was still little more than an outline when the company's newest operation began to move wood.

To superintend the Grays River operation, Corman hired Otto Keiski, a rangy, bespectacled, forty-seven-year-old logger from Naselle. Otto had the nervous energy of three men, a low boiling point, and a remarkable memory for names and faces—the perfect formula for a woods boss. Corman and Keiski selected capable woodsmen to fill other top posts, including John Haglund as logging foreman, Dave Woods as engineer, and Bill Woods as construction boss. In the office, Bob Vining kept the books, and all forestry-related matters fell to Don Borglum.

Keiski asked Walter Paavola to serve as bull buck to oversee the cutting crew, which at first numbered only eight men. He offered Walt's son Bobby, fresh out of Naselle High School, a job on the cutting crew too. The younger Paavola soon found himself on the business end of a clunky Disston chainsaw. "Otto said we'd get lazy if we got the smaller saws," Bob said with a chuckle years later. His first partner was Charlie Wirkkala, who was later killed in an accident while working for Crown Zellerbach. Other notable members of the original cutting crew included Vern Larson and Eldryd "Slim" Crenshaw, both of whom would later become bull bucks.

Otto Keiski, superintendent of Weyerhaeuser's Grays River operation from 1955 until 1965. Photo courtesy of Keith Metcalf

The logging plan called for the cutters to start in Section 21. This square mile contained one of the finest stands of Douglas fir The Almighty ever planted, which later tallies quantified as 80 million board feet. That figures out to be 125,000 board feet per acre, twice the volume the average acre in Grays River yielded. On September 15, 1955, fallers Vern Larson and Burl Broyhill encountered one of the largest Douglas fir that would come out of Grays River. The tree measured twelve feet two inches across the stump and stretched nearly the length of a football field. The butt cut, a twenty-four-footer, scaled 11,300 board feet and contained enough wood to construct a five-room house. It was such a spectacular specimen that the entire crew posed for a picture in front of it.

The first logging crew was composed mostly of former employees of the Wirkkala Brothers and Deep River Logging Company, which after a half-century of logging in the Naselle country had recently closed its camp. The equipment at their disposal was Wirkkala Brothers' machinery—a couple of sled-mounted yarders and five Allis-Chalmers dozers. Being gentle ground, most of Section 21 was logged with "cats," the loggers' all-encompassing name for crawler tractors. High climber "Trembly" Bob Tremble rigged one high-lead spar

When a spar tree could not be found, loggers put one up in the proper position. This photo from the spring of 1962 shows a Weyerhaeuser crew raising a spar tree in the upper Grays River country. Vern Larson photo courtesy of Nancy Larson

Curt Anderson (top) and Matt Varila pose on a thirteen-and-a-half-foot spruce they felled near Blaney Lookout in the fall of 1966. Vern Larson photo courtesy of Nancy Larson

tree, however, to the base of which came seven million board feet of the big round stuff. A 1951 Washington TL-15 Trakloader, the Wirkkala Brothers' most modern piece of machinery, stayed mighty busy swinging logs onto trucks. When it could not keep up with production, the giant cuts of fir were rolled onto trucks with cats using a two-tiered system of loading ramps.

Wirkkala Brothers' old truck garage and blacksmith shop in Weyerhaeuser's Grays River camp, 1965. Photo courtesy of Keith Metcalf

Weyerhaeuser did not have trucks of its own at Grays River when logging commenced in 1955. Contractors R. V. Larson and Earl Spargo got the job. Rudolph "Duffy" Larson, already a two-decade veteran behind the wheel, put four trucks on haul. Earl Spargo, whose small fleet had been hauling logs for Deep River Logging Company since 1939, sent five trucks to Grays River. Normally, all of the trucks made four trips to the dump at Deep River daily. The logging works slowed some during a recession in 1957, but when the outlook improved, Weyerhaeuser transferred ten big rigs from Longview to Grays River. After the yellow and green fleet arrived, Larson continued to haul for Weyerhaeuser while Spargo opened a truck dealership in Longview.

One of four yarder sides operating out of Weyerhaeuser's Grays River camp in 1971. Vern Larson photo courtesy of Nancy Larson

From these humble beginnings, the Grays River operation grew steadily. In 1958, there were two high-lead sides running full blast when Fred Waara signed on as hook tender. Waara was a no-nonsense, old school logger. Though only a little above average height, he was as burly as a man could be. He typically spoke in a soft, grumbly tone, but when he took occasion to flex his muscular voice, young choker setters were known to have broken into tears. Fred Waara even intimidated Don Pearson, a robust specimen himself, when they began working together in 1959. "I had the utmost respect for him," Pearson recalled after a half-century in the woods. "He was one of the best loggers I ever worked with."

The expanding logging operation called for an ambitious road-building plan. The Wirkkala Brothers' original access road up Grays River became the arterial of the ever-expanding network. It was eventually designated the 7000 Line and every

Weyerhaeuser's logging plan called for moving its loggers to lower elevations during the winter months, but that did not always work out. During the late 1960s, there were some significant snow events that brought the operation to a standstill. Vern Larson photo courtesy of Nancy Larson

A truckload of Grays River timber goes over the "brow log" and into Deep River in 1965. In the 1970s, Weyerhaeuser expanded its log dump at Svenson's Corner to include a dry land sorting facility. Photo courtesy of Keith Metcalf

road branching from it given a sequential number. Throughout the late 1950s and early 1960s, construction crews extended the 7000 Line into the high country along a broad ridge leading to the 2,700-foot summit of K. O. Peak. They also encircled the forest to the east of Grays River between Blaney Creek and Mitchell Creek with roads numbered 7200 and 7300. A succession of engineers laid out these right-of-ways, including Dave Woods, Bert Chamberlain, Billy Vaughn, and Boyd McCall. Ed Siepelt was the last construction supervisor in Grays River, that position falling under the engineering department around 1962.

Although known as Weyerhaeuser's Grays River operation, the company had considerable holdings in the Salmon Creek, upper Naselle, and Bear River Ridge areas. By 1962, road construction crews had completed the 8000 Line to the top of the ridge separating Bear River from Dell Creek, just west of Naselle. Being nearer the coast, the timber in the Dell Creek block ran heavy toward hemlock and cedar, yielding about forty thousand board feet to the acre. The density of the timber mattered little when Typhoon Frieda blew ashore on Friday, October 12, 1962. Its hurricane force winds that the region would long remember as the Columbus Day Storm leveled mile upon mile of forestland. "We were working right on top of the [Dell Creek] hill when the Columbus Day Storm came," noted timber faller Bob Paavola. "Oh, we had a mess there."

All told, the Columbus Day Storm leveled some eighty-three thousand acres of Weyerhaeuser forestland in Oregon and Washington. Logging plans were drastically revised, and, like the Yacolt Burn sixty years earlier, a massive salvage operation began. Much of the recovered timber was hemlock, which at the time was fetching low prices in domestic markets. Japanese buyers, however, welcomed the soft, white wood. From the company's docks at Aberdeen and Longview, Weyerhaeuser sent its choicest logs across the Pacific to Japan, Korea, and later China. Exporting raw logs became a lucrative venture for Weyerhaeuser throughout the remainder of the twentieth century.

Back in Grays River, tragedies of a human nature struck the operation in the mid-1960s. First, on June 26, 1965, Otto Keiski was involved in an automobile accident that took the life of his wife, Tekla, and left him a paraplegic. A month later, Robert Chamberlain, an eighteen-year-old recently graduated from Naselle High School, was killed while working on a road construction crew. Chamberlain had the grim distinction of being the first fatality at Weyerhaeuser's Grays River operation; he would not be the last. On Saturday, September 23, 1967, a choker

Woody Gifford

A number of noteworthy occurrences took place within Weyerhaeuser Timber Company in 1959. For one thing, it dropped its middle name, Timber, and became Weyerhaeuser Company. It also adopted its well-known corporate logo of a green triangle framing the outline of a tree. Also in 1959, the name Woodrow J. Gifford was added to the firm's payroll ledger for the first time.

Woody Gifford and his wife, Julia (Mannikko), moved to Deep River in 1940. He worked a little in the woods but spent most of his time farming. Then he became a bucker for Weyerhaeuser. In the mid-1960s, Woody began contributing colorful poems and character studies to a new publication called *Loggers World*. His poetry was also featured in several books put out by Loggers World Inc. In 1974, he retired from Weyerhaeuser and moved to Seaview. Four years later, a collection of his poems went to press in a chapbook titled *Timber Bind, Logger-Rhythms of the Great Northwest*. Finley Hays, founder of *Loggers World*, wagered: "If Woody Gifford is not the best poet to come from the woods, he ranks right among them!"

Norm Penttila, a chainman on Weyerhaeuser's engineering crew, lends perspective to the great girth of the western red cedar that grew in the Dell Creek and Bear River areas in this 1972 photograph.

cable parted sending a log crashing down upon sixty-two-year-old logging foreman John Haglund in the rugged country around K. O. Peak. Nearby and scarcely a week later, a choker setter trying to avoid a steep downhill hike slid down a cable and was impaled on a jagged stump top. All told, nine workmen lost their lives in logging-related accidents at the camp.

Bob Harvey, an Oregon State University graduate, became superintendent at Grays River after Otto Keiski's departure. Since the operation started in 1955, wooden spar trees had been the standard logging setup, but under Superintendent Harvey, the transition to portable steel towers, which began in 1964, continued. The new towers were faster and less complicated to rig up than their wooden counterparts and became something of a necessity in Weyerhaeuser's coastal holdings where trees suitable for spars were mighty scarce. By 1968 the change-over was well underway when Harvey introduced grapple logging to the Grays River operation.

In 1967, George H. Weyerhaeuser, the great-grandson of the company's founder, unveiled a forest management program called High Yield Forestry. It was the culmination of many years' work by company researchers to grow trees "faster, straighter, and stronger." The four basic components of High Yield Forestry were improved seed stock, rapid regeneration (i.e., hand-planting), fertilization, and thinning. Weyerhaeuser had long prided itself as being a leader in the tree-growing field, and this strategy set the bar high—its aim was to produce forty-thousand board feet of merchantable timber per acre in forty years time. To open up room for the vigorous, young growth, with its ever-increasing volume, Weyerhaeuser upped its harvest of the "decadent" and "over-mature" old growth, which was losing volume through decay. This made the late 1960s and 1970s a banner period for logging at Grays River.

High Yield Forestry was still in its infancy when Delbert "Pinky" Anderson, a former Wirkkala Brothers logger, became the superintendent at Grays River in 1969. At the time, six yarder sides could be found scattered around the operation, as well as two cat sides. Longtime shovel operator Marvin "Marvy" Badger recalled that his most productive day loading trucks came around this time while at the levers of the old Washington TL-15. With two bulldozers feeding the landing on the 7260, Marvy sent forty-seven loads to the dump in eight hours. Pinky Anderson's tenure as head push was short-lived. In 1971, he and his wife, Dorothy, moved to Borneo where he supervised a Weyerhaeuser logging operation. Three years later, he was knifed by a disgruntled Bornean employee and while recuperating suffered a stroke that ended his working days.

The same year Pinky Anderson assumed the superintendent's role at Grays River, Keith Metcalf replaced Boyd McCall as chief engineer. Metcalf began working at Grays River in 1961 and had a pretty thorough knowledge of the area when he became the head road architect. It was a good thing, too, because the increased logging called for the construction of between twelve and twenty miles of road per year. Some of those roads went into territory new to the operation. To better balance its harvest goals, in 1965 Weyerhaeuser transferred nine thousand acres of old growth fir from the Pe Ell branch to Grays River, bringing the camp's managed lands up to eighty thousand acres. Across that spread, Metcalf, Party Chief Russell Wirkkala, and the other members of the company's engineering crew worked steady to keep well-designed roads out in front of the loggers.

One of the special projects that Metcalf oversaw took shape in 1974 on a large flat between Alder Creek and the main stream of Grays River. There, Weyerhaeuser constructed the Alder Creek Fish Pond to raise Chinook and coho salmon. The facility was built, maintained, and staffed entirely with funds provided by the timber company. A nearby salmon hatchery furnished the fingerlings, which were reared in the pond before being released into the stream. According to Metcalf, prior to Weyerhaeuser's fish-rearing efforts, salmon runs did not exist above the Grays River Falls. It was only after the 1950s, when a large charge of dynamite reduced the cataract to a cascade, that spawning salmon even had a chance to ascend beyond the Grays River Gorge. For nearly a decade, Weyerhaeuser-raised salmon smolts began their life's journey from the Alder Creek Pond.

Having been part of Weyerhaeuser's Longview Division from the start, in the late 1960s Grays River became part of the Twin Harbors Division. In the years to follow, personnel from the Raymond office began filling positions at Grays River, including logging foreman Milton Noyes and cutting foreman Ellis

Working some 180 feet above the ground, Fred Waara prepares to rig one of the last wooden spar trees used at Weyerhaeuser's Grays River operation in the mid-1970s. Photo courtesy of Keith Metcalf

Ruddell. The switch also created new destinations for company and gyppo truck drivers. Besides hauling to the dump at Deep River and the occasional trip to Longview, logs from the Grays River operation were now sent to Weyerhaeuser facilities in Raymond, Cosmopolis, and Aberdeen.

In 1971, towering Bill Rehder became the last superintendent at Grays River. A basketball standout at the University of Washington, Rehder came to the operation after serving as a shop foreman at Weyerhaeuser's Road 12 Camp, part of the Longview Division. The first few years of the Rehder era witnessed some notable milestones. Around the time Fred Waara strapped on his climbing spurs and belt to rig the last wooden spar tree, Weyerhaeuser crews began cutting the first major stands of second-growth timber. Large portions of the operation's holdings had been logged early in the century and, by the mid-1970s, were once again covered with merchantable trees. In 1975, new faces and equipment arrived from Molalla, Oregon, after Weyerhaeuser phased out its operations on the Santiam River after thirty years. Logging continued at a steady rate even as the last substantial stands of old growth were cut in the late 1970s. In addition to company sides, Weyerhaeuser engaged a number of contractors, including Grays River Logging Company and Durrah & Martin Logging Company, both of which spent considerable time working in the rugged, cedar-covered hills around Dell Creek and Bear River.

With the 1980s came a nationwide recession, the likes of which the timber industry had not experienced in a half-century. The decade began with the eruption of Mount St. Helens, which leveled some sixty-eight thousand acres of prime Weyerhaeuser forestland. Cutting crews from Grays River were transferred to the mountain to assist in the salvage effort. The loss of timber in the blast zone, however, was a mere hiccup compared to the massive backslide in overall business conditions. The company began to streamline. In

By the late 1970s, Weyerhaeuser's old growth reserves in Grays River were dwindling. Here timber fallers Jim Raff and Kayse Rogers pose by a massive fir, circa 1978.

1982, Weyerhaeuser executives started trimming unprofitable diversification efforts and shuttering noncompetitive processing facilities. Over the next three years, Weyerhaeuser slashed its payroll by nearly ten thousand.

In February of 1983, Weyerhaeuser's reorganization effort struck Grays River. The tasks of cleaning up the logging jobs and camp continued through the summer of 1983, when the last Weyerhaeuser employees at Grays River were let go. Some of the loggers found work at the company's Cosmopolis and Longview branches, but most of those who remained in the area went to work for gyppo outfits. In 1984, Keith Metcalf, who had been working out of Weyerhaeuser's Aberdeen office, returned to manage the lands as a contract-timber sale area. In 1988, Hancock Forest Resources, a timber investment management organization and subsidiary of the John Hancock insurance company, purchased the forestland—all eighty thousand acres. Within a decade, gyppo log trucks loaded with second-growth logs were rolling down the 7000 Line out of the country still called Weyerhaeuser.

"The Climber Was Away"

by Woody Gifford

Auda Grier. Photo from Walt Busse Collection

Local logger, rancher, and poet Woody Gifford lived in Deep River for thirty-four years and worked at the Weyerhaeuser Grays River operations. Known to be a kind, observant man, his poems were often published in Loggers World. *There are different theories on the woman who inspired "Peggy" in the poem. Some believe it was Peggy Raemhild Busse, who set chokers in Deep River in the forties, while others think it was based on an actual incident with Auda Spencer Grier, as related by her son, Norman. "Mom worked for Knappton Logging in 1942. They were short on loggers, so she was asked to be a whistle punk. One day the mainline broke. The regular climber wasn't there, so Mom said she'd do it—and she did!"*

Back in the "Warring Forties" on the Brix Bros. Logging Show,
They was runnin' might short-handed but still tryin' to make her go;
The boss was 'a stormin' and frettin' a beggin' The Draft Board for men,
They promised relief was a comin' but Red Tape could never say when.

Well, the boss made another wide circle—he checked all the jungles and bars,
It cost him a pocket of shekels and a box or two of cigars.

At breakfast time the next morning, while The Boys were munchin' away,
The hooker groaned—he was short a punk and couldn't whistle that day;
Peggy, the hasher, came trippin' by with a platter of bacon and eggs,
Then Shady remarked, with a devilish grin as he glanced at her well-shaped legs,
"Hell, Oscar, here's Peg—let her whistle a few—she'll put some speed in those Yegs."

When The Crew went out to the crummy and started to climb aboard,
There sat Peggy, caulk boot and all as calm and relaxed as a Lord;
Well, Shady looked kinda sheepish awright, course the Crew couldn't help but grin,
The Crummy went out to The Loggin' Show and The Loggin' started in.

The hooker found out with no room for doubt that Peggy was sharp on that wire,
Course The Boys tried to mix her up a bit, but nary a signal got by her.

Now they'd logged halfway 'round over broken up ground,
And were gettin' ready to swing;
Twas a Blue Monday dawn—and the Climber was gone,
He'd just felt the urge of spring;
Well the hooker was gettin' excited—He was sure in the bight alright,
So he asked The Crew if someone could climb, but they all sat quiet
and tight.
When Peggy had finished coilin' her wire, she tripped right on up to
the tree,
Says Peg, "If you Panty Waists can't climb, step aside, won't you please
and let me!"
The Boys looked around and down at the ground, then they quietly
eased away,
Peggy lashed on her spurs and buckled her belt, then she pulled the
Pass Line for play,
She adjusted the sling—took a hold of the ring—then she said, "Jim,
please take her away."

Peggy Raemhild Busse. Photo from Walt Busse Collection

Jim was nervous as hell, cause a shapely young belle had ne'er tightened his pass line before,
He throttled down slow—but Peg signaled—go! Go! And to the top of that tree she did soar;
Peggy called out a hoh!—to Jim down below, then she swung her rope 'round the tree,
She jabbed her sharp spur in that Douglas fir, and proceeded to let herself free;
Now she circled The Tree with the end that was free of that pass line—then chained up the yoke,
She moved 'round to clear, the block, strap and gear, glanced toward the ground—then she spoke;
"Up easy, Jim, ho! Thank you, now go—Up a bit, that's just ducky, now ho!"
That cussed block slid, 'round those tree plates and grid just like it knew where to go;
Peggy unhooked the choke of that chain 'round the yoke—then she made herself fast on the line,
"Up—easy, Jim, ho!" She let the rope go—"Now down, Jim, please, all the way down."

We were all kinda' nervous, as Jim let her down from her perch way up there on that tree,
She was cuter than hell—and she did her job well but it kinda' did something to me;
As Peg gently touched down, she sorta' glanced 'round at The Boys then she tossed her brown curls,
"Now don't you think—she said with a wink—this separates The Boys from The Girls?"

CHAPTER 6

The Gyppo Loggers

by Bryan Penttila

The word *gyppo* is unique to the logging industry. No one is quite sure from where the term derived, but through the years it has come to describe an independent logging firm that harvests timber under contract for a large landholder. Small logging outfits have been a mainstay in the woods of the Willapa Hills since well before the word *gyppo* ever crossed a logger's lips, but it was not until around the time of the Second World War that the modern incarnation of the gyppo logger appeared on the scene.

Skamokawa was a hotbed for early log trucking operations. This image from the summer of 1927 shows Iverson & Winters' hard-tired truck with a one-log load heading down East Valley. Photo courtesy of Kent and Irene Martin

By the time of the Great Depression, technological advancements had given loggers new tools to harvest trees. Chief among these were the bulldozer and the log truck. With plank or rock-ballasted roads, the gyppo loggers found work aplenty in isolated stands of old growth that the big railroad logging companies bypassed and in the lowlands where sawlog-sized timber was now growing on sites that had been logged a half-century before by bull-team and early steam operations. The combination of relatively inexpensive mobile machinery, handy timber, and expanding markets made possible this new breed of logger.

Gyppos were, and still are, an individualistic breed, notable for their ingenuity, stubbornness, and restless work ethic. Their outfits came in all shapes and sizes and might be as small as single owner-operator or a complex enterprise with one hundred or more employees. No matter the scope of their business, logging was a way of life for the gyppos, for their work did not end with the yarder's quitting whistle. Many gyppos went broke, the luckier ones eked out a comfortable living, and a tiny percentage prospered to become leading figures in the region's timber industry. The buzz of countless logging outfits echoed through the Willapa Hills during the twentieth century, and though it is impossible to detail every operation, what follows is a brief description of some of the more notable gyppo loggers.

This Johnson Brothers Logging Company plank road was nearly identical to the wooden causeway Anton and Walter Brix built into the Bear River watershed. There they maintained a small sawmill to cut hemlock planks to surface the road. Photo courtesy of Jim and Darrylin Eastham

Brix Logging Company, 1934–1958

The Brix name had long been associated with the old Columbia River sawmill town of Knappton when Anton Brix appeared on the scene in 1934. Anton's older brother, P. J., had owned the mill and the adjoining settlement since 1909. The younger Brix had recently acquired a tract of forestland over the hill in Bear River and, to harvest the stand, formed Brix Logging Company with his son Walter. To access the timber, the Brixes built a plank road. The course resembled a railroad, but instead of being topped with steel rails, rough-cut four-by-twelve-inch hemlock boards laid abreast created a surface for the trucks to drive on. Henry Pellervo began his logging career laying timbers on the road in 1942. "It was an awful job,"

Brix loggers in Ellsworth during the 1940s. Left to right: Johnny Bighill, Art Paavola, and Arvin Simukka.

he remembered with a chuckle. "The biggest job we had was to jack the log trucks back onto the plank road. They'd be slipping off every now and then."

Among the most notable events that occurred at the Brix operation took place in 1943 while most of the able-bodied men were off serving their country. On this particular day, the mainline parted and came whizzing through the bull block, piling up in a terrific heap at the bottom of the spar tree. No one volunteered to ascend the spar and re-thread the bull block. Then Auda Grier, the forty-five-year-old wife of the company's bookkeeper and mother of three, stepped forward. This wartime whistle punk was lashed securely to the pass line and hoisted up the 180-foot pole. Mrs. Grier performed admirably high atop the spar tree and for her deed earned an everlasting place in loggers' lore as "the lady high climber."

By 1942, Walter Brix was itching for more independence. He formed Walter G. Brix Logging Company and took to the woods behind Chinook and, later, around Grayland, Washington. He then moved to Briceland, California, where he engaged in sawmilling and logging before he succumbed to a heart attack in

1954. Meanwhile, Anton continued to manage the operation at Knappton. In 1943, he moved the outfit to Ellsworth Creek on the lower Naselle River, where he secured a section of state timber that was decimated by a hemlock looper infestation in 1931. The company's two high-lead sides highballed through the state wood, and before long they had a contract to harvest adjoining stands of Weyerhaeuser timber.

In 1944, as Anton prepared to retire, he made plans to sell the company to Erwin Anderson, one of his young employees. Brix saw in the youth a capable successor and, beyond that, appreciated Anderson's worldview. In 1938, Anderson, still tipsy from his most recent drinking spree, heard Orson Welles' broadcast of "War of the Worlds" and thought Armageddon was upon him. That night he gave his soul to Jesus. Anton Brix, a one-time Methodist preacher himself, never got the chance to retire. He died of a heart attack shortly after approaching Anderson with the offer. Brix's widow saw to it that Anton's wishes were carried out, and Anderson became the owner of Brix Logging Company.

An oversized spruce log parked at the Nemah crossroads. Rueben Penttila stands on top of the log and lowboy driver Stanley Stenholm on the truck. Photo courtesy of Philip Penttila

"Because of the war, men were very difficult to get and hard to find," Anderson said in 1998. "It was tough. I learned to bulldoze, and I had to high climb." But he managed to carry on and in 1951 took on his old friend Rueben Penttila as a partner. Seven years later, they dissolved Brix Logging Company. Anderson then had several stillborn sawmill ventures in both British Columbia and Washington State. He bounced back and soon found himself bound for the Philippines where he managed various American-owned logging operations. Erwin and his wife, Ellen, spent the next seventeen years in the mahogany forests of the Philippines where they became co-owners in a sprawling plywood plant and sawmill. He retired in 1985, and the couple moved to Seattle where they spent the remainder of their lives.

Penttila Logging Company, 1958–1977

"Rueben was a logger. Erwin, he did his logging in the office." That is how Mike Ullakko, a longtime Brix Logging Company employee, described the differing approaches Rueben Penttila and Erwin Anderson took toward the logging business. Although they did not become partners until 1951, Penttila and Anderson had tramped around the area's logging camps together since the Great Depression. When their company was underbid on a large tract in the Cruiser Creek drainage of Nemah in 1958, the duo parted and Rueben founded his own company.

Hook tender Curt Gray and Rueben Penttila recline on Penttila Logging Company's Skagit BU-80 yarder in the 1960s.

From the former Brix operation, Penttila Logging Company assumed a log loader, a couple of dozers, and a trio of gas-powered, mustard-colored Mack trucks. Many of the Brix loggers carried over with Penttila, too, including Charles "Hoopy" Holm, who started working for the company in 1951. Holm cut timber, topped and rigged spar trees, drove truck, and by 1955 was manipulating the levers of the company's log loader; a log loader would be Hoopy's office for the next forty years. Penttila Logging Company's first yarder was a triple drum unit mounted on an Allis-Chalmers HD-14 that was rented from Francis Ziak, a Knappa logger who had recently relocated to Naselle.

Crown Zellerbach provided steady work for Penttila. His crews built roads along Jim Crow Creek and at Brookfield developed Crown's log dump facility in 1958. He upgraded to a new triple drum yarder and put Howard Laney at its controls. The single high-lead side stayed busy year-round, while cat sides skidded logs for Crown and other landowners seasonally. More often than not, Penttila Logging Company's operations were at the western edge of Crown Zellerbach's holdings in the hills around Nemah and back of Chinook.

The 1960s were a good time for Penttila's company. To his lineup of trucks, he added three new Macks, all a fabulous shade of red, in 1963. Rueben's fondness for red also carried over to his yarder. From Skagit, he special-ordered a new yarder, a BU-80 high-lead machine, the whole works painted a rosy hue. This big unit became the centerpiece of the outfit. Under its 120-foot steel spar could be found Penttila's longtime employees like Wally Raemhild, Mervin "Phatt" Bighill, George "Cy" Robinson, Bill Nitzel, and Don Pearson. Curt Gray, another woods veteran, kept busy as the company's multiskilled cat skinner. When the situation demanded it, the old triple drum, which by the mid-1960s was sporting a steel tower of its own, was rigged up as a second side and the crew divvied up to man it.

Being a small operation, Rueben treated his crew like family, and some of them were. Penttila Log Cutters, a partnership between Rueben's brothers Paul and Herbert, Gordon Benson, and Virgil Guggisberg, handled his timber cutting. His equipment was maintained

Right: No history of logging in the Naselle area would be complete without mention of the L & M Garage. From 1958 until 1982, brothers-in-law Millard Penttila (shaking hands with a Union Oil representative) and Leonard Vaughn serviced all types of logging equipment. "Laying in the snow under a Cat with water running down the back of your neck wasn't something you looked forward to," Leonard remembered, "but it was our job, so we did it."

by the mechanics at the L & M Garage, owned by Rueben's youngest brother, Millard, and Millard's brother-in-law, Leonard Vaughn. Curt Gray, too, was Rueben's brother-in-law. But even those who were not direct relations found Rueben to be even-tempered and always willing to offer advice.

Rueben dedicated much of his time to civic affairs. He served for a good many years as chairman of the Naselle School Board and, when the new high school was built, donated his machinery and time to construct the football field. He helped found The Bank of the Pacific in 1971, which opened a branch not far from his home in Naselle three years later. In 1977, Rueben sold Penttila Logging Company to an up-and-coming logger from Cathlamet named Jerry DeBriae. Rueben Penttila passed away a decade later at seventy-three.

Virgil Guggisberg, a partner with Gordon Benson and Paul and Herbert Penttila in Penttila Log Cutters, stands at the butt of a large spruce tree. The firm cut timber under contract with Penttila Logging Company, owned by Paul and Herbert's brother Rueben. Photo courtesy of Philip Penttila

Durrah & Martin Logging Company, 1946–1994

Logging came naturally to the Durrahs. Their father, Elmer, was an enterprising logger in Eden Valley before he settled into the life of a dairy farmer. It was on the family farm that the Elmers' sons—Harold, George, and Jim—learned early and sometimes-unpleasant lessons about hard work. Harold, the oldest sibling, had made some money gillnetting and put it down on a farm in the Elochoman River Valley. It just so happened that the acreage contained a stand of yellow fir. Harold, his brother-in-law Alex McCormick, and younger brother Jim got a notion to turn the trees into cash. They enlisted the help of their other brother, George, who was newly married and working as an electrician. Together they formed Durrah & McCormick Logging Company in 1946.

Ken Martin and George Durrah, together with George's brother, Jimmy, owned Durrah & Martin Logging Company from 1965 until 1983. The pair is seen here at a logging job near Randle, Washington, in 1983, around the time Martin retired from the firm. Photo courtesy of Kent and Irene Martin

"We started with nothing," Jim said of the venture. He owned a worn-out dump truck that he converted into a log hauler, and the new company bought an old gas-powered Best 60 bulldozer. "We got it for junk price, and George made it run." The brothers' ingenuity also overcame the fact that the machine did not have a winch, and with it they both skidded and loaded the logs. From Harold's farm, the company cut a few other small farmer patches around Cathlamet and Skamokawa before taking a contract to log for Crown Zellerbach in 1948. Harold, who did not feel suited for a life in the woods, left the company around this time to pursue a career as a commercial fisherman.

Meanwhile, in 1947, a pair of Skamokawa siblings named Ken and Harold Martin began logging together with a war surplus TD-18 bulldozer. Like the Durrahs, the Martins were go-getters and grew Martin Logging Company through hard work and long hours. By the 1950s, the brothers were opening

Contract trucker Gene Brock stands next to his "homemade" log hauler, a 1943 Reo with a Geiberson war surplus engine, around 1957. Brock was among a large number of independent truck drivers in the area that included M. F. "Mickey" Paulsen, Harvey "Peck" Larson, and Gilbert "Ecky" Erickson. Photo courtesy of Kent and Irene Martin

up a vast stand of old growth fir for Crown Zellerbach in the upper reaches of Skamokawa's East Valley. On one of the sheer slopes in that rugged country, Harold rolled a bulldozer and was killed in 1959. Ken carried on and together with Harold's widow, Ida, ran the outfit six years more before the companies merged to create Durrah, McCormick & Martin.

The company's main log-getter was a sled-mounted BU-135 Skagit yarder. It had been built as a high-lead machine, but when the partners started to revive the long lost art of slackline logging, they had a third drum added. Later, a tank retriever trailer replaced its wooden sled and a steel tower rounded out the unit. In 1965, George's son, Russell, began working for the company. At the time they were gyppoing for Weyerhaeuser along the Toutle River, not far from Mount St. Helens. Around this time, the partners modernized with the purchase of a new BU-95 Skagit with a 110-foot steel tower. Jim took charge of the BU-95 side, George ran the BU-135 side, and Ken, as he modestly put it, was the company's "bull cook." A few years after purchasing the big Skagit, they purchased a bigger Skookum-Tyee yarder that became George's pride and joy.

When Alex McCormick sold his shares in the company in 1967, the business became Durrah & Martin Logging Company. After spending a number of years along the Toutle, by the early 1970s, Durrah & Martin put its two yarders to work closer to home, both for Weyerhaeuser and Crown Zellerbach. After the 1980 eruption of Mount St. Helens, the company joined the mad dash to salvage the downed timber in the blast zone. Not long after the Mount St. Helens salvage operation ended, Ken Martin retired. As the last of the region's old growth was logged, the partnership, now Durrah Logging Company, traded its oversized slackline machines in for smaller models. In 1994, George, who was seventy-three, and Jim, who was sixty-eight, retired, just shy of a half-century of gyppo logging.

Grays River Logging Company, 1940–1983

If even half of the tales about Jim Oatfield are true, it is little wonder that he became a legend in his own time. Stories abound of the resourceful tactics he employed to keep his loggers productive and his log trucks rolling to the dump. Jim's "mighty will" was even celebrated in a poem by noted logger-poet Woodrow "Woody" Gifford. Oatfield's tenure in the woods began in 1924 when he was just sixteen, and in his near-sixty-year logging career that followed, Jim's cork shoes tread the far reaches of the Willapa Hills. For much of that time he was the majority owner and driving force of Grays River Logging Company.

Jim Oatfield, co-owner and woods boss of Grays River Logging Company.

Theodore Swanson, Arthur Anderson, and Alfred Peterson founded Grays River Logging Company in 1940 to harvest a tract of timber near Pillar Rock. Swanson, the son of a local pioneer family, began logging during the bull team era and since 1908 had been the owner of the Grays River Shingle Company. The partner's second job took them to Skamokawa where things had hardly got going before a mishap sent their steam donkey barreling into the river. "All you could see of the donkey," recalled Swanson, "was about six inches of the exhaust pipe." They fished the engine from the river, however, and continued logging.

Grays River Log grew, and in 1948, when Alfred Peterson passed away, Jim Oatfield assumed one-third ownership. Upon Swanson's retirement in 1961, his shares in the company were divided between Oatfield, Swanson's daughter, Gloria Clark, his son, Teddy, and his son-in-law, Gene Whitten. Each of the company's co-owners served a function with the firm: Oatfield managed the logging; Gloria handled the books; Whitten supervised log loading; and Teddy focused on road construction and later drove the company's lowboy. During this time, Oatfield's sons, Richard and Raymond, became increasingly active in the company as well. Also key to the outfit were high-lead foreman Walt Smith, rig-up boss Karl Johnson, and cat skinner extraordinaire John Erickson.

By 1970, Grays River employed over 115 men and operated seven high-lead sides. In the years to follow, the number of high-lead sides grew until Grays River Log was in the running with Ben Thomas Logging Company of Woodland as the biggest gyppo operation in southwest Washington. The rapid growth did not come without some pains, and by the mid-1970s, Grays River had scaled back to four yarder sides. When he was not supervising the moving or setting-up of a yarder, Oatfield could often be found at the controls of a cat. He was a masterful cat skinner and given half a chance would yank a good portion of the logs off a yarder setting before the crew even had a chance to rig up.

That log-hungry drive was evident in Oatfield throughout his career. The trait was not hard to spot in his sons, either. In 1971, Richard and Ray started the successful Willapa Logging Company in Raymond, and later hired their stocking-capped brother David, who had also spent years with Grays River Log. In 1980, seventy-two-year-old Jim Oatfield was still out in the woods every day overseeing his fifty-five employees and multiple subcontractors. But even the legendary gyppo could not keep up this pace, and his health began to fail. The venerable Grays River Logging Company was dissolved and on October 12, 1983, its equipment liquidated in a public auction across the highway from the Naselle High School. Jim Oatfield died three days later.

Johnson Brothers Logging Company, 1925–1957

Brothers Charlie, John, and Alex Johnson grew up immersed in logging. Their father, Andrew Wirkkala, began rolling logs into the Naselle River shortly after he arrived in the valley in 1883 and changed his name to Johnson. Their oldest brother, Matt, was also an independent logger. Then, on a fateful Monday in 1912,

Moving Johnson Brothers' double drum loader on a makeshift trailer in the 1940s. Left to right: Alvin Larson, unknown, John Johnson, co-owner of the outfit, and Matt Saari, the loader engineer. Photo courtesy of Jim and Darrylin Eastham

a cable parted as Matt's crew navigated his new steam donkey over a steep stretch of ground above the Naselle River falls. Down the mountain the donkey went, uprooting trees in its path, before it piled up at the foot of the hill one thousand feet below. The operation limped along for another year, but Matt was whipped. A line from the *Willapa Harbor Pilot* in June of 1913 summed up the situation: "The Matt Johnson camp came to a sudden closing last week, and on Monday all of the loggers, including the lady cook, went to South Bend to file liens on the remaining logs for their wages."

High climber Chester "Buster" Raemhild, donkey puncher George Waara, and Johnson Brothers co-owner John Johnson after a hard day's work. Photo courtesy of Jim and Darrylin Eastham

The younger Johnsons faired better. By the 1920s, they were trucking logs out of the South Nemah country on plank roads and down the new Pacific Coast Highway to Johnson's Landing, the brothers' log dump. About a mile west of where Naselle High School stands, the Johnson boys built a camp. The site was home to a truck shop, office, cookhouse, and several large bunkhouses. These bunkhouses provided living quarters for the numerous transient loggers the Johnsons imported from Raymond and Aberdeen to supplement their local workforce. Logging matters fell to woods boss John Johnson, while Charlie dealt mostly with the business end of the operation, and Alex kept the camp running smoothly.

In 1934, Reino Johnson, the owners' nephew, hired on to punk whistle for the outfit. Inspectors were quick to note that he was only fifteen years old and, according to newly instituted child labor laws, Reino could not work in the woods for another year. With their nephew taking a year off to ripen, the Johnsons gave another Naselle teen his first job in the woods. His name was Fred Warra. Fred would become a familiar face at Johnson Brothers and other area operations for the next forty-five years.

With the Second World War came a crippling shortage of men and supplies that forced the Johnson Brothers to close their camp for a spell. Once they were up and running again, Johnson Brothers' loggers worked steadily along the lower Naselle River, particularly on the south slope of the Radar Hill, where they built two small lakes that have become popular recreation spots. While visiting Finland in 1952, Charlie Johnson died unexpectedly. He was fifty-eight. John and Alex kept the business running, though it was not always a harmonious undertaking. On occasions when their tempers flared, they would stand back-to-back and conduct their business meetings in booming voices that could be heard echoing through camp. The system worked, and Johnson Brothers Logging Company carried on until Alex's passing in 1957, at which point John closed out the business and retired.

R. V. Larson Trucking, 1940–1980

Hauling logs has been in the Larson family's blood for well over a century. Victor Larson, the family's patriarch, married a fellow Swede named Ida Swanson and the young couple settled into Brix's logging camp above Sisson Creek. That was in 1908, and the following year their first son, Rudolph, was born. "Duffy," as the little Larson was nicknamed, was an active and inquisitive child. When the family moved to Portland Lumber Company's camp at the summit of KM Mountain, Duffy was a regular passenger in the cab of the Shay locomotive with his engineer-father. "I knew as much about a steam locomotive as kids know about cars today," Duffy said in 1994. As a teenager, he put that knowledge to good use when he ran the forty-two-ton Shay while crews picked up the rails on the old KM line.

Roy Nelson, co-owner of Larson & Nelson Trucking, checks the wrappers on his load while parked along the 5800 Line in Salmon Creek, 1991.

By the 1930s, truck logging was coming into vogue. Duffy fit well behind the wheel of a big rig and by 1940 was the owner-operator of a log truck. Within a few years, he owned three trucks. Much of the time, Duffy's small fleet was kept busy hauling logs for Grays River Logging Company where his uncle Theodore

Swanson ran the show. Another truck working for Grays River at the time was Willard Hyrkas' red and black, single-axle International. When Hyrkas was not available to drive, a Salmon Creek teenager named Roy Nelson filled in. The Larson and Nelson families were united in 1948 when Duffy's daughter, Delores, became Mrs. Roy Nelson.

Nelson went to work for Grays River Logging Company for a spell, but in 1956 returned to work for his father-in-law, who was busy hauling logs for Weyerhaeuser. In 1958, as the log trucking business was taking off, Duffy, his brother Harry, son Dennis, and Roy Nelson opened Oneida Ready-Mix on Deep River. Being the sole cement plant

Wayne Bond Logging Company's crew about 1958. Front row, left to right: Harold Morse and Bob Haataja. Second row: Russell Anderson and Reino Anderson. Back row: Billy Wuorinen, Jim "Gator" Gatens, Paul Deaton, Kenny Edmunson, Stanley Rangila, Ted Deaton, and Wally Anderson. Wayne Bond was a Naselle-area gyppo logger for Crown Zellerbach from 1956 to 1963. Photo courtesy of Stanley Rangila

in the area, Oneida Ready-Mix's two trucks hustled to haul its product to the ever-expanding Air Force base west of Naselle. Before the company was dissolved in 1967, it was responsible for pouring countless local house foundations and even brought slurry to construction sites as far as Kelso.

At its peak, R. V. Larson Trucking had a dozen big rigs on the road. They shuttled logs for Weyerhaeuser, Crown Zellerbach, Grays River Logging Company, Durrah & Martin Logging Company, and numerous other gyppo operations. In 1980 when Duffy retired, his sons Dennis and John formed Larson & Nelson Trucking with Roy Nelson. They maintained a fleet of seven trucks, but when Weyerhaeuser shuttered its Grays River operation in 1983, they began to scale back. Finding new hauls took the drivers far afield. "We were chasing stumps all over the country," John Larson said of that turbulent period. But Larson & Nelson endured, and in 2011 it is not unusual to pass a truck on the highway with the Larson name painted on the door.

Schmand Logging Company, 1936–1959

Though he was the son of the village blacksmith, the ring of hammers on an anvil did not appeal to Casper "Cap" Schmand. Logging was more his tempo. Cap was nine years old when the family settled in Grays River in 1912 and, within a few years, was working in the woods. He hung up his cork shoes long enough to attend Oregon State College but was soon back in the hills above Grays River. He worked close to home, saved his money, and during the Great Depression started Schmand Logging Company.

In 1931, Cap Schmand bought a used donkey and rigged it up on a show along lower Grays River. Arthur Harrison and Delwin Larson cut the timber, and Harvey and Peter Larson bucked the trees into logs. A few years later, he incorporated Schmand Logging Company. He logged near Clatskanie, Oregon, and was working the hills back of Skamokawa by the outbreak of the Second World War. His outfit moved to Salmon Creek for a spell, and by 1943 Schmand's yarders were chugging logs out of the canyons along Hull Creek. Crown Zellerbach put him to work on Old Smokey, where his two high-lead sides plugged away for many years.

Cap married Bhradde Mae Masson in 1934, and together they had five daughters and a son. On Sunday, September 5, 1945, Cap took his young daughters to his jobsite. While their father tinkered with some equipment, the kids busied themselves playing on a pile of logs. Without warning, the deck began to move. The children scrambled for safety, except eight-year-old Sandra Lee, who was overtaken by a log and crushed. Charles Crippen, who was working with Cap that day, had the sickening task of driving the grief-stricken father back home to Grays River with his lifeless child in his arms.

For a number of years, Jim Oatfield was Schmand's right-hand man. Oatfield kept the outfit productive until he went to work for Grays River Logging Company where he eventually became a majority owner. Ed Engels replaced Jim Oatfield as the company's foreman. By the 1950s, Louis Penttila and Toivo Kari handled Schmand's cutting on a contract basis. Two of their top workmen were Alfred Bailey and Floyd Mattson, a hulking cutter known to handle a two-man chainsaw by himself. Deep River's Eino Rangila served as bull buck. For a

One of Cap Schmand's log trucks gets loaded with a single cut of fir.

time, Schmand even branched out to harvesting trees in British Columbia with the Vancouver-based Beacon Hill Timber Company. Cap selected his old friend Homer Bjornsgard to head to Canada and run the show, which lasted only a few years.

By the late 1950s, Cap was tiring of the logging game. The ever-increasing competition among the gyppos had served to lower logging prices and Schmand had had enough. He had a large and productive farm on Hull Creek to tend to and was active in a host of civic activities. Among his many notable community achievements was helping co-found the Grays River Volunteer Fire Department and serving as chairman of the Naselle-Grays River Valley School Board. In 1974, he was named Old-Time Logger of the Year for Wahkiakum County—an honor he rightly deserved. Casper "Cap" Schmand passed way August 10, 1988. He was eighty-five.

Valley Eight Logging Company, 1946–1982

Among the many partnerships to go gyppo logging after the Second World War, few were as colorful as Valley Eight. The octet that united to form the company were all former employees of Deep River Logging Company and residents of the Salmon Creek Valley. John "Buddy" Bjork, Oren Hyrkas, Archie Larson, Walter Leikas, and Elmer Tienhaara had recently mustered out of the military in 1946 when they joined Waino Kilponen, Armas "Spike" Lampi, and Gunnar Nelson to become Valley Eight. None of the partners had much in the way of financing, but in Eden Valley they purchased an affordable gas-powered donkey from logger Bob Manthe and picked up a dozer at a war surplus sale. Most of the partners owned farms along Salmon Creek, and that is where they turned for their jobs.

Five of Valley Eight Logging Company's founders, circa 1948. Left to right: Walt Leikas, Gunnar Nelson, Armas "Spike" Lampi, Waino Kilponen, and Oren Hyrkas. The three partners not shown were John "Buddy" Bjork, Archie Larson, and Elmer Tienhaara.

Valley Eight started buying and logging timber in Salmon Creek and, before long, was logging under contract for Crown Zellerbach. By then, Elmer Tienhaara had sold his stake in the outfit leaving the seven remaining founders to carry on. Gunnar Nelson worked as hook tender before settling into the role of master mechanic at the company's shop on the Kilponen farm; Waino Kilponen was Valley Eight's cat skinner par excellence; Archie Larson ran the yarder and later the company's log loader; Walt Leikas handled the high climbing and worked in the rigging with Spike Lampi and Buddy Bjork; Oren Hyrkas was a landing man and spent much of his time as head loader. Joining Valley Eight's principals were Waino's brother, Eugene, and Oren's brother, Willard, who drove the company's bright red log trucks.

Those who saw him work rated Waino Kilponen as one of the most skillful cat skinners anywhere. This photo shows him ripping through a stump with the "stump splitter" attached to the rear of his International TD-25 in Bear River, 1973.

The company's logs went into Deep River, first at the Johnson Brothers' dump near the highway bridge, and later, at Peters & McKinnon rollway, just off Miller Point Road. There the sure-footed Lassila brothers, Henry and Oscar, assembled the rafts. "Them guys could be so drunk that they could barely crawl out of a car," Gunnar's son Roy Nelson remembered. "And they'd go down the ladder off the dump and get on the boom sticks. They could walk clean to the end of them and never fall off." Insobriety at Valley Eight was not limited to the booming grounds. By some accounts, there were more bottles of beer—usually quart-sized containers known locally as "Valley Eight stubbies"—hoisted at the company's shop than in either of the local taverns.

In the mid-1950s, Liekas and Kilponen traveled to southeast Alaska to look into the prospects of working for the U.S. Forest Service. The Forest Service was opening up portions of the Tongass National

Finn Valley Cutters was a contract timber falling firm started by Norm Penttila (left) and Kenneth "Bear" Tyyska in the 1990s. Also in this 1999 photo are Ryan Wirkkala (center) and Lyle Haataja (right).

Forest to large-scale logging, and with the formation of the Ketchikan Pulp Company in 1953, new markets were available for the wood. Valley Eight's owners decided to pass on the chance to expand, but another Naselle logger, Joe Penttila, capitalized on the opportunity. In 1956, he and his wife, Dorothy (Grier), and their three young daughters moved to Petersburg and a short time later started Penttila Logging Company on Kuiu Island, where the couple would operate successful logging camps for nearly three decades.

In 1960, Valley Eight ended its large-scale logging activities. The partners sold their big yarder, Gunnar Nelson retired, and several of the founders sold out. Kilponen and Leikas carried on with a cat side and began to focus more and more on road building. The pair, along with Waino's son, Dale, built countless miles of road for both Weyerhaeuser and Crown Zellerbach. On the last day of 1982, Waino Kilponen and Walter Leikas dissolved the company.

Martin Nygaard Logging Company

Martin Nygaard Logging Company at work on Longfellow Hill above Naselle in 1993.

With the completion of the Astoria-Megler Bridge in the mid-1960s, a well-established gyppo from Astoria began working on the north shore. The company's founder and namesake was Martin Nygaard, who had grown up working on the Skipanon River log boom owned by his father and uncle. Martin busied himself on the boom through his college years. He graduated from Oregon State University in 1951 and continued to work on the log boom. In 1958 he went logging for himself. Soon, his company's time sheets were filled with the names of notable Knappa- and Astoria-area logging families like Barendse, Flues, and Thomason. His foreman was Robert "Kewpie" Ziak, a barrel-chested logger known for getting logs and spinning yarns.

Nygaard sold his logging equipment to Dant and Russell in the mid-1970s and became the logging manager for the company and was on the board of directors. When Dant and Russell went broke in the early 1980s, he purchased the equipment back and resumed logging on his own. In the meantime, his sons, David and John, had started an outfit called Nygaard Brothers Logging. In 1989 the family consolidated its holdings and formed Warrenton Fiber Company and set to work on an expansive sorting yard and log chipping facility along with their logging and road building operations.

Martin Nygaard with a brand new D8T Cat in Warrenton, Oregon, 2011. Photo by Bertroch

D8T
NYGAARD LOGGING
WARRENTON, OREGON
CAT
PETERSON CAT
www.petersoncat.com

CHAPTER 7

Wirkkalas in the Woods

by Bryan Penttila

No name looms larger over the woods of the Willapa Hills than Wirkkala. The surname represents multiple branches of a clan whose origins can be traced back to the town of Kaustinen in west-central Finland. As a general rule, the original immigrant Wirkkala families bore numerous offspring. Some of the children were born in Finland, others after the family arrived in America, but nearly all became loggers. So did many of the third generation. By the mid-twentieth century, virtually every logging outfit in the area had at least one Wirkkala on its payroll at one time or another. The woods were full of them. Finley Hays, the founder of *Loggers World* magazine, once quipped that his publication "had many subscribers with the last name of Wirkkala. More Wirkkalas than Smiths, Joneses, or Andersons."

No less than a half-dozen branches of the Wirkkala family tree landed in the Naselle-Grays River area. The earliest to plant roots in Naselle were brothers John, Charles, and Andrew, who settled on adjoining homesteads along the upper reaches of the river. Although Andrew changed his name to Johnson, his sons—Matt, Frank, Alex, John, Emil, and Charles—ranged the hills with axes and saws for more than a half-century. These Johnson brothers are treated elsewhere in this book. The two scions of the family that contributed most substantially to the area's logging history were the descendants of distant cousins Abraham and Andrew Wirkkala. Abraham's five sons all became loggers, including Oscar, the most famous woodsman to carry the family's name. Andrew sired a dozen boys who all worked in the woods, and several of them owned successful logging enterprises.

The Abraham Wirkkala family arrived in Washington State at various times in the late 1890s, and he and his older boys caught their first whiffs of Douglas fir while working in a Grays Harbor sawmill. In 1899, the family relocated to Deep River and began logging. Abraham and his sons—Andrew, Oscar, Matt, Sam, and Alex—bought a small steam donkey and put it to good use skidding logs for Finnish farmers in Deep River and Naselle. Around 1906, the Wirkkalas moved to Quincy, on the Oregon shore. There they encountered some mighty steep terrain that taxed the limits of ground-lead logging. Oscar had a notion as to how to speed things up, though. In 1908, he went to Portland, had a carriage built from a simple drawing he had made, and when he returned to Quincy, strung a skyline cable between two ridges and began flying logs out of the canyon with an early version of slackline logging.

Logging jobs took the Wirkkala brothers all over western Washington. Andrew, the oldest son, married and settled in Clatskanie where he logged, farmed, and preached at a nearby Apostolic Lutheran Church. Matt became a lifelong logger, as did Sam who died in a logging camp in the Philippines in 1929, leaving a wife and six children. The youngest sibling, Alex, and his wife, Hilma (Hansen), settled in Seal River in the 1930s and raised a passel of logger-sons, including Bill, Walt, Ernest, Eddie, Art, George, and Charlie, and a daughter, Irene. In the 1950s, a timber falling accident left Walt permanently disabled, and in 1961, his brother Charlie was killed while cutting timber for Crown Zellerbach.

Oscar Wirkkala, at left, around 1915, stands on the four-drum donkey he designed and Washington Iron Works built. Oscar was the most famous of the Wirkkalas in the woods and, over his career, patented numerous innovations within the logging industry. Photo courtesy of Ernie and Ed Wirkkala

Ed Wirkkala puts the "back cut" in a big Douglas fir in the Grays River country, circa 1977. Weyerhaeuser bull bucks often teamed new timber fallers with Ed so that they could learn to down the big sticks the right way.

Three of Charlie's sons—Allan, Lloyd, and Orval—have become notable figures in the logging world. Allan, the oldest, began driving for Valley Eight Logging Company in 1960 and, after gaining experience with a good number of gyppo outfits, settled in with David Howell's Lone Pine Logging Company where he became foreman. In the early 1980s, he started Twin Firs Logging Company and later Allen Wirkkala Logging Company. Lloyd, too, began driving log truck after his school days, even though he was only seventeen years old at the time. He spent some time gyppo logging and in the mid-1980s formed L. W. Enterprises, a trucking firm that at one time included four big rigs. Many of the logs that Lloyd's trucks hauled came from Liberty Logging Company, a Castle Rock–based firm founded by his brother Orval in 1986.

Orval, Lloyd, and Allan Wirkkala. Photo courtesy of Lloyd Wirkkala

Of the five sons of Abraham, none became more prominent in logging circles than Oscar. Over the years, his inventive mind led to a score of U.S. patents, including improved skyline systems, slackline carriages, donkey engines, blocks, and marine propellers. Perhaps his most practical invention came in 1922. According to family lore, Oscar was laying in the Whatcom County clink at Wickersham when he took particular note of how the jailer's key fit neatly into the lock. That observation let to the Wirkkala Choker Hook, better known as the "choker bell," which ever since has been the principal means by which loggers fasten cables to logs. Oscar's ideas were so revolutionary that he spent much of his time in court protecting his patents. His extended legal battle with Donnis Bardon over his Choker Hook was one such example. Wirkkala won some royalty fees in 1935, but by then foundries across the Northwest were turning out choker bells on his pattern. Oscar died in 1959 and, a couple of decades later, was among one hundred people inducted into Washington State Historical Society's Centennial Hall of Fame.

Patented by Oscar Wirkkala in 1922, variations of the Wirkkala Choker Hook are still in use today.

In 1906, Andrew and Hilda (Korpela) Wirkkala

High climber Oiva Wirkkala circa 1940. Photo courtesy of Daryl Wirkkala

left their little cabin at Brix Brothers' logging camp for their newly acquired "stump ranch" on the South Fork of the Naselle. To this couple were born twelve sons—Arne, Oiva, Howard, Ed, Lauri, Albert, Hannes, Martin, Fred, Ben, Paul, and Urho—and three daughters—Elizabith, Eva, and Amy—between 1904 and 1927. In the Wirkkala household, everyone contributed to meeting the necessities of life. As soon as they were physically able, the boys went to work in the fields and forest while the girls lent a hand with domestic chores. Arne worked in the local logging camps until he died following a motorcycle accident in 1928. Oiva left home as a young teen and tramped around the big logging camps from Nemah to Nehalem. Howard, Ed, Martin, and perhaps some of their brothers found their first jobs at the Brix Brothers' camp a few miles south of the family farm.

Their upbringing instilled deep bonds between the siblings and created a pack of strong-willed woodsmen. The family's cooperative spirit was evident in their first logging show. It was in 1937, and ten of the eleven brothers went in together to take a logging contact offered them by the Johnson Brothers Logging Company to harvest a half-section of timber belonging to South Bend dentist Rudolph Schenk. Oiva, who had learned the high climbers' trade while working on the famous Kerry Line railroad in Oregon, rigged the spar trees, while Howard was designated foreman.

With Schenk's timber logged, in 1939, Martin, together with his brother-in-law, Oiva Johnson, began logging the Hundis family property in Naselle under the title Wirkkala & Johnson. A few years later, Wirkkala & Johnson built a truck shop a quarter-mile west of the present Naselle High School that became the company's base as its crews worked their way up Cowan Peak. In 1940, Albert, Hannes, Ben, and Paul organized Wirkkala

Brothers Logging Company and set about cutting big fir in the Naselle area, paying a dollar per thousand for stumpage. Despite the lingering effects of the Great Depression, both companies got off to a good start.

In 1937, ten of the eleven Wirkkala brothers teamed up to log a half-section of timber south of Naselle that belonged to South Bend dentist Rudolph Schenk. Urho, the youngest sibling, was not quite old enough to take an active part in the operation. Photo courtesy of Wally Wirkkala

America's entry into the Second World War forced changes upon the Wirkkalas' operations. In 1941, the U.S. Air Force sent a draft notice to Oiva Johnson, who spent the next several years stationed in the South Pacific. In the meantime, Martin took on another partner named Johnson. His new associate was Alfred "Al" Johnson, no relation to Oiva, and a member of an Astoria family that operated a successful grocery business. Lean and hard working, Al quickly proved himself a capable logger. Hannes received a draft notice from the U.S. Army in 1942, which reduced the sibling count at Wirkkala Brothers by one, and a short time later, Fred and Urho were drafted into the service as well.

A big fir at Wirkkala Brothers Logging Company in 1945. Left to right: Homer Durrah, Paul Wirkkala, Gus Magnuson (at end of log), Matt Saari, Pastor Niilo Sepala, Wayne Keiski, Bill Wirkkala, unknown, Archie Laine, and Wilbert Salme.

Wirkkala & Johnson's Clyde yarder during World War II. Left to right: Al Johnson, Ward Pakenen, Lauri Wirkkala, Harvey Johnson Sr., Martin Wirkkala, unknown. Photo courtesy of Brenda Wirkkala-See

Salmon Creek Logging Company co-owner Urho Wirkkala sits at the controls of a new Allis-Chalmers HD-21 as Earl Anderson and Fred Wirkkala (far right) look on. Urho built hundreds of miles of logging road during his career, worked on the Burma Road during World War II, and even put roads into the jungles of Nicaragua for the Evans Products Company in the 1970s. Photo courtesy of Gary Wirkkala

In 1945, the Wirkkala Brothers jumped their operation over Salme Hill and began harvesting Henry Wilme's timber in Deep River. A semiretired logger, Wilme punked whistle for the outfit. It was while logging his timber in the fall of that year that Wilme was killed when the rigging sent a tree crashing upon him. Despite the loss, the Wirkkalas finished clearing the Wilme tract and relocated once again, this time on the Left Fork of Grays River.

When Hannes, Fred, and Urho returned home from the service, they joined with Lauri to incorporate Salmon Creek Logging Company. The company got logs out of the Naselle area until the early 1950s when its crews began to open up a stand on the Left Fork of Grays River. Hannes, ever restless with the urge to do the Lord's work, sold his shares in the company and became a full-time preacher, later settling with his wife and family in Brush Prairie, Washington. Among the remaining partners, Lauri served as donkey puncher, Fred ran the rigging crew, and Urho became head cat skinner.

Around 1952, Martin purchased Al Johnson's shares in the company and re-incorporated the firm as M. W. Logging Company. Despite the fact that Martin's worldviews put him at loggerheads with the local union representative, Herb Faubion, M. W. developed a large and stable corps of employees, including Bernard "Bim" Raff, Chester "Buster" Raemhild, Severi Wiitala, Joe Barnett, and Milton "Moke" Bighill, among many others. In the early 1950s, Martin hired a husky lad from Long Beach named Sid Snyder, who, after a few months of setting chokers, realized that logging was not his calling and went on to open a grocery store in Seaview. Overseeing the old hands and greenhorns was Martin's brother Howard, the logging superintendent, reckoned by some to be the savviest logger among the siblings.

The first half of the 1950s was a busy time for Wirkkala Brothers Logging Company. After the war, the brothers began to open up the big fir country along the upper Grays River. They put two high-lead sides in a state sale and hired the best high climber around to top and rig their spar trees. It was their brother Oiva. Besides climbing, Oiva dabbled in salvage logging with Ed, who had returned to the woods after spending time on the evangelical circuit. Then there was a falling-out between Oiva and some of his siblings, so he left the area and spent the next decade topping trees in the redwood country around the Oregon-California border. In 1955, Albert, Ben, and Paul sold their camp, equipment, timber, and log dump to Weyerhaeuser Timber Company in what *The Lumberman* magazine reported to be three-quarters of a million dollar deal.

With the proceeds from the sale, the three brothers pursued various investments. Albert spent the remainder of his life engaged in evangelical work, though on occasion he returned to logging. One of his later ventures was Flintstone Logging Company. Around 1968, Albert bought the well-worn equipment of a gyppo who had gone broke while logging along the Elochoman River just west of Cathlamet. With a crew composed largely of his siblings and their sons, Albert logged "Flintstone Swamp" before dissolving the concern in 1970. Paul partnered with Buck Wisner in a Buick dealership in Longview and invested in several hotels. Ben pursued oil exploration in Texas and South Dakota. He also devoted time to his love of flying and built a landing strip in his backyard. In the late 1950s, Ben hired Jesse Salme, Arvin Simmuka, Arnie Jarvi, and Stanley Rangila to clear and build Wirkkala Airport.

In the mid-1950s, Salmon Creek Logging Company became associated with Aberdeen Plywood through the sale of its peeler grade logs. The plywood manufacturer persuaded Lauri, Fred, and Urho to move their operation to Cougar, Washington. Around 1956, the brothers, along with most of their crew, moved into the newly assembled bunkhouses at the company's camp along the Lewis River. It was a busy place at the time, as Swift Creek Dam was taking shape, but through the hubbub Salmon Creek logged for Aberdeen Plywood and soon had Forest Service contracts of its own. By 1961, Lauri, tired of living and working so far from home, pulled out of the outfit and returned to the family farm where he continued to work on-and-off as a logger.

By the time Lauri settled back in Naselle, his brother Martin had focused his energies on the Naselle Rock Company. In 1945, Martin bought a quarry along Lane Creek and over the years used his loggers' downtime to develop the operation. By the late 1950s, Naselle Rock was a going concern. M. W. Logging Company had its last major show in a patch of state timber on Naselle Ridge in 1960, but by then Martin had banked his future on rock. It was a wise decision, and a decade later, he installed a state-of-the-art asphalt plant at the quarry and his firm became the Naselle Rock & Asphalt Company, which remains one of the area's largest private employers.

Wirkkala Brothers' high climber Oiva Wirkkala and log truck driver Delbert "Pinky" Anderson in 1954. Albert Wirkkala photo, courtesy of Wally Wirkkala

Even as Martin phased out of the logging game, a younger generation of Wirkkalas was getting a feel for the business. In 1960, Howard's son, Howard Jr. "Sonny," started South Fork Logging Company and began thinning for Crown Zellerbach in the Nemah bottomlands. Thinning blossomed into clear-cut logging after the Columbus Day Storm of 1962. Then Weyerhaeuser called. In 1963, he re-incorporated as Howard Wirkkala Logging Company and entered the high-lead game north of Raymond. He picked up a used yarder and a new log loader, manned by his brother-in-law Gordon Chapman. Louisiana-Pacific gave him his next job. "We got good production. We had Junior [Floyd Madsen] and Matt [Varila] and everybody out there . . . We could get 20 loads a day, easy. Things were happening fast."

Lauri's sons, Vic and Lorne, formed V & L Wirkkala Logging in 1964 and began subcontracting for Howard Wirkkala Logging. Soon they put their dozer to work thinning for Crown Zellerbach. That lasted about five years. In the early 1970s, Vic and Lorne reunited once again and took scarifying jobs, cutting contracts, and, for a time, even operated a portable sawmill. In the early 1970s, their younger brother, Gary, hired on with Crown Zellerbach and three decades later retired from the pulp mill at Wauna. Lauri's youngest son, Phillip, traded in his teaching job for falling timber and spent his working days making stumps.

By 1968, after working in the mountains around Cougar and Carson for a decade, Salmon Creek Logging Company began bidding on Forest Service sales around Port Angeles. For their new show, Fred and Urho invested in a mammoth Washington yarder that came with an equally large price tag. The firm worked around Port Angeles for scarcely a year before a string of setbacks sent Salmon Creek into a spiral. Bankruptcy followed close behind. Fred took the loss hard and moved back to Naselle a broken man. Urho teamed with Howard to build roads in the Cascade Mountains for Weyerhaeuser and the Forest Service, and in 1973 became construction foreman for Mayr Brothers of Hoquiam. Urho's son, Daryl, followed in his father's hobnailed footsteps and has spent his career moving logs.

In the meantime, gyppo jobs had taken Howard Wirkkala Logging Company all across western Washington, from Quinault to Sultan to Stevenson. Like so many other gyppo loggers, Sonny experienced financial reversals but always managed to bounce back. His last big high-lead show came in the Cascade Mountains above Packwood in the late 1980s. The job came to an abrupt end, however, when the Forest Service curbed timber sales in the district. Sonny showed his resilience once again and, in the early 1990s, started The Land Company with his cousin Lorne. Together they bought forested properties and logged them in true gyppo fashion. As the twentieth century came to a close, The Land Company and Russell Wirkkala Forest Service, a construction business owned by Ben's son, stood as the last logging enterprises conducted by the descendants of Andrew and Hilda Wirkkala.

Howard "Sonny" Wirkkala Jr., with a block and strap on his shoulders in 1979, while logging north of Hoquiam.

Deep River was home to other noteworthy logging Wirkkalas. Charles F. and Ellen Wirkkala lived in the center of the community where Charlie's parents, Nestor and Hulda, ran Deep River's Shamrock Hotel, considered by many to be among the finest boardinghouses on the lower Columbia. Ellen eventually became the proprietor of the Shamrock, while Charlie worked right behind it as the log dump and boom manager for Deep River Logging Company. He held that post from 1929 to 1956. Charlie's son, Kenny, who is equally well known as "Buckshot," grew up in a house adjacent to the hotel and watched daily as the big sticks were rolled off the rail cars and met the water with a terrific splash.

It is little wonder that Kenny's first job was on the boom. Before long, he was setting chokers for gyppo loggers like Jimmy Lindros and Clyde Sprague. Early in 1964, Kenny formed a company and went logging for Crown Zellerbach, which he would continue to do for eighteen years. One of Buckshot's friends and favorite employees was Bobby Wirkkala, the grandson of Naselle's first John Wirkkala. In the early 1980s, when Kenny's sons, Kevin and Loren, graduated from high school, they too looked to the woods for work. Kevin bought his own log truck, and they both worked "on the outside" for a time before joining with their father in Ken Wirkkala & Sons Logging Company. Both Kevin and Loren settled in Cathlamet where they are interested in a number of businesses. Ken retired in 2004.

Lorne Wirkkala—timber faller, cat skinner, and co-owner of The Land Company of Naselle—in 2002.

Just up the valley from the Shamrock Hotel was the farm of Henry and Martha (Jurselius) Wirkkala. Their sons—Willard, Melvin, Richard, Arnold, Leslie, and Lester—all became loggers, though Lester died as the result of a gun handling mishap in 1948. Leslie and Willard both towered over six feet five inches, possessed full frames, and could have been easily mistaken for Paul Bunyan. They were timber fallers, first at Deep River Logging Company and later as contractors. Melvin injured his back in a logging accident and spent the remainder of his career in the Longview Fibre paper mill. Richard was a top-notch truck driver and hauled logs for several gyppos including Ray Burton Logging Company. Arnold, too, worked for Ray Burton, though his laboring days began at age thirteen in a pair of cork shoes he borrowed from a neighbor. Except for Melvin who lived in Longview, Henry Wirkkala's sons all made their homes in either Skamokawa or Cathlamet.

As the first decade of the new century dawned, there was only a handful of Wirkkalas still working in the woods. The older generation had either passed away or retired, and their children found logging to be an increasingly undesirable career choice. The younger Wirkkalas were not alone. Stagnated wages, long commutes, and unpredictable employment have sent the descendants of many old-time timber families into new lines of work. Chances are good, however, that as long as people are cutting trees and hauling logs, there will be Wirkkalas in the woods for generations to come.

CHAPTER 8

Old-Time Loggers

by Donna Gatens-Klint

I dedicate my chapter to my beloved brother, Pat Gatens, who was tragically killed in June 1979 falling timber for Vickaryous Timber near Megler, and to my son-in-law, Randy Oman. These two men were my favorite loggers of all time.

Pat Gatens

Pat Arthur Gatens

Pat was a logger and a fisherman. He went to work right after high school working for Weyerhaeuser in Grays River. In later years, he worked for different timber contractors.

He loved to make people laugh. He cut wood for many of the older people in the community. If they had money to pay, he charged them. If not, he would give it to them. Pat grew up in Nemah, and spent many hours in boats and around our father's oyster company. Pat and Starrlette Smith were married in 1962. They had three children.

Randy Oman

John Randal "Randy" Oman

Randy was my son-in-law and the father of my two oldest grandchildren. Randy loved logging more than anyone I've ever known. There wasn't a thing he didn't like about it. Randy left us too soon in life at age forty-seven, due to cancer. Randy worked for Bob Olsen Logging most of his career, he and Bob closer than brothers. Bob says he can never be replaced as he did everything perfectly.

Randy could be a real comedian. One day, he had friends over. After dinner, Randy set the plates on the floor so the dog could lick off the plates, then he picked up the plates and put them in the cupboard. The company sat with their mouths open. Of course it was all a joke.

Randy was born in 1961 and died in 2008. He had two children who meant the world to him.

Stanley Rangila

Stanley Rangila. Photo by Diana Jordan

The summer Stanley was thirteen, he went to work for Fred Johnson in Deep River and got hooked on logging. Stanley worked for different loggers in his career. He ran a cat and loader for Ben Wirkkala when Ben cleared land that became the Wirkkala Airport. He also worked for Cap Schmand, Weyerhaeuser, and Wayne Bond, then Crown Zellerbach from 1964 to 1985, then Durrah & Martin for eight years. In later years before retiring, he worked for North Fork Logging from Chehalis, Hamilton Logging, and Russell Durrah.

Stanley Rangila was born to Eino and Martha Rangila in1937 in Centralia. A second-generation Finn, he has lived in Deep River his entire life. Stanley married Helen Simukka in 1959. They have two children and four grandchildren. Today Stanley restores old logging equipment and farm machinery.

William "Billy" Wuorinen

Billy Wuorinen. Photo by Lisa Bighill

Billy was born in 1930 in South Bend, the only child of Victor "Vic" and Laura Anderson Wuorinen. Billy grew up in Toonerville till he was nine years old when his family moved to Naselle. Billy married Midge Treichler in 1958. They had two children. Midge passed away in 2007. Bill went to work for Wirkkala & Johnson in the 1940s when he was seventeen years old, working on Radar Ridge. Billy was a head rigger for many years; then he hook-tended and climbed from 1947 to 1963. He worked for Crown Zellerbach till they shut down in 1985. He cut for a few gyppos before his retirement. Billy has always loved woodworking, making beautiful clocks, table lamps, and burl tables. As a lover of history, Bill has donated much of his logging equipment collection to the Appelo Archives Center.

Gilbert "Gib" Haataia

Gib Haataia

Gib started logging right out of high school, working for Grays River Logging. After getting hurt in a car wreck, and being off work for several months, he then went to work for Crown Zellerbach. He stayed with Crown for twenty-seven years, except for the time he was in the service. For Crown, he started off planting trees, then moved on to the rigging. He also drove dump truck, finishing off his career at Crown driving log truck for many years. After Crown shut down, Gilbert worked for Jerry Debriae for nine years, and Benson Logging from Oregon for several years. Other employers included Kevin Wirkkala and Harvey Johnson. He finished his career working for Russell Wirkkala.

Gilbert was born in 1939, and graduated from high school in 1957. He married Sarah Lorton in 1960. They have three sons. Dan Cothren remembers that, "Gib loaded the trucks at Deep River. He had the first hydraulic grapple rig. Gib was really good at it, too. I worked under Gib at Crown for fourteen years. He was really good to work with."

Robert "Bobby" Saari

Bobby Saari, Astoria, 2001

Bobby went to work for Weyerhaeuser in 1972 at nineteen. He continued to work for them until they closed their operations in 1983. Bobby then went to work for the alder mill in South Bend. He decided to go fall timber for different gyppos, including Don Fudge, Steve Wirkkala, Finn Valley Cutting, and Joe Helvey. Bobby is now self-employed. He climbs, tops trees, and does tree removal. In the summer and early fall, he travels with the Great Lakes Timber Show from Escanabi, Michigan. In the show he does topping, speed climbing, logrolling, and other events. His sixteen-year-old daughter, Emily, also performs in the logging show. Bobby and his wife, Becky Johnson, were married in 1976 and have six daughters.

Walt Busse. Photo by Diana Jordan

Walt Busse

Walt and Eddie Busse were batching it one time. The rest of the family was gone. They were neglecting things at home and were out drinking until the wee hours of the morning. They didn't build a fire when they returned home, even though it was very cold and freezing out, and even colder upstairs where they slept. Walt had a pair of super thick logger's pants called tin pants. When they got wet they could stand up by themselves. When Walt got up in the morning, he had a big gnarly hangover. It was so freezing in the house. He stumbled around looking for his tin pants so he could go to work. For him, you did not ever miss work unless you were dying. Walt could see frost on those pants, but they were all he had. As he put on the pants, he realized they were frozen solid from being so wet and cold. Off to work Walt went wearing a big hangover and frozen pants. Walt was miserable all day. From then on he decided to behave and take care of business at home.

Frank Vollmer

Frank Vollmer

When he was a young boy, Frank went to Star of The Sea School in Astoria. One day when he was outside, a nun, Sister Angela Carmelle, asked him what he planned to do in life. He asked her if she could see the big tugboat down on the river and told her, "I plan to be captain of a boat like that one day." Sure enough, that dream came true. Frank worked on the Columbia River for forty-eight years, first as a deck hand for eleven years, and then as captain of various tugboats for thirty-seven years. He towed log rafts from Deep River to the sawmills in Cathlamet, Longview, Warrenton, and other places on the Columbia River. For his last five working years, he ran small tugs to transport bar pilots and river pilots out of Astoria to their ships.

Rodney "Rod" Smith

Rod Smith

Rod went to work for Weyerhaeuser in 1959 on his eighteenth birthday, working there for four years. He then went on Crown Zellerbach's cutting crew for a couple years. He drove log truck for Carl Skirvin and Kenny Wirkkala, and then bought his own dump truck. Weyerhaeuser needed more dump trucks, so he ended up owning six trucks. Rod bought a cat from Waino Kilponen, and rocked roads. He hauled with the off-road trucks for twelve years until Weyerhaeuser shut down their Grays River operations. Rod then converted one of his dump trucks into a self-loader and hauled piling materials. Rod owned and operated his own timber company GRT (Grays River Timber). He then bought into C&J Logging, and worked there a few years. His son, Steve, took over the trucking business, and Rod retired in 2003. Rod and Cecile Axmaker were married in 1960. Their second son, Christopher "Kit," passed away in 1987. They also have two grandchildren.

Robert "Bob" William Olsen

Bob Olsen, left, and Brandon Olsen, right

Bob started as a gyppo logger in 1985. He decided he would stay small with only one employee, Randy Oman. Randy ran loader and Bobby ran cat. Bob feels "the main thing that's changed in logging is the cost of operating. Not only has the price of fuel and equipment gone up but also the state's Labor and Industries rates. The L&I rate for anyone running a power saw is more than most people make per hour." He contract-logged for International Paper from 1985 to 1994 in the Castle Rock area. One summer when he was logging at Mount Rainier, the crew all lived in travel trailers. "That was a lot of fun," he commented.

Bob now logs with his twenty-four-year-old son, Brandon. He likes to work for private landowners now because he's working for individuals. Bob's father was a timber faller, and his brother, Billy, also works in the logging industry. His uncle is Jerry Debriae. Bob married Donna Giding, and they have two children and a granddaughter.

George Edward Klint

George Klint

George went to work in the summers at the age of twelve working in the woods for Johnny and Harry Kapron making one dollar an hour. He thought he was really getting rich! At seventeen, he lied about his age and went to work for Durrah & McCormick setting chokers. He also worked for Cap Schmand and drove truck for Ray Mooer Logging, Peters and McKinnon Logging, McNally, and Grays River Log. When working for Grays River Log, he hauled from Long Island. The Island was an interesting place as there was so much wildlife—bear, elk, deer, and coyotes. In later years, he bought his own log truck and hauled for Joe Florek, Durrah & Martin, and Valley Eight.

George loved to play tricks on the guys in the woods. He would nail their lunch pails to a stump. When they picked them up, the handles would come off, or the bottom would pull apart, but the guys usually got even with him one way or another. He ate Gilbert Magnuson's lunch one day. Gilbert didn't say anything to him, but was laying for him the next day. Gilbert had his wife bake a cake and frost it with chocolate laxative.

George was born in 1934 and grew up in Grays River milking cows. His father died when he was six years old, and the kids had a lot of responsibility helping their mother keep the farm going. In 1958, George married Donna Gatens. They have two daughters and three grandchildren.

Lyle Haataja

Lyle Haataja. Photo by Karen Bertroch

Lyle graduated at seventeen from high school in 1965 but didn't start working in the woods till he was eighteen. He set chokers and fire-watched for Weyerhaeuser in the summers while going to college. He then went to work for Ray Burton. After four years in the Navy, he went back to Burton choker setting and working as second rigger. He then went to Alaska to log, where he lived in the logging camps. The only way in or out was by a single engine seaplane. He worked eleven seasons in Alaska. In the winter months, he would come home and work for the local loggers. Lyle did a lot of high climbing and timber falling for Durrah & Martin, Norm Penttila, Kenny Tyyska, Steve Wirkkala, and he got hurt working for Joe Helvey. Lyle and his wife Loanne have five children.

Dale Rose, left, and John Footh, right, at the Loggers Reunion at the Archives Center, January 2011. Photo by Diana Jordan

John Footh

In 1954, John went to work for Weyerhaeuser and was there twenty years until they closed down in 1984. He then went to work for Chuck Torppa and Johnny Bergquist for eight years. Until Weyerhaeuser shut down their logging operations, he chased and ran cat and loader. At C&J, he ran a yarder.

John married Frieda Olson in 1963. They have three sons, Johnny, Robbie, and Paul who has two daughters. John is their sitter. He takes care of them when they get off the school bus and he goes bike riding with them. He says, "It keeps him in shape."

Merlin Durrah

Merlin Durrah

Merlin started his logging career fire watching in the summers when he was seventeen years old. In 1958 when out of high school, he worked for Campbell Brothers Logging for a short while. Merlin reports, "safety was always the same when he worked in the woods." The big change he's seen in the logging industry was the change from the size of the logs that went from old growth to small timber.

His favorite truck was a brand new 1979 Mac B Series. Another big change was from the wooden spar tree to the towers. He said the towers were much better. Instead of taking three days to rig, it only took about four hours. Merlin only worked for a few companies. He drove for Brock for two years, and worked for Helvey Logging Company second loading and head loading. Most of his logging career was for Crown Zellerbach from 1965 to 1985. He worked in the shop and on the construction crew, and drove truck. Merlin was in a couple of log truck wrecks, but neither was serious.

Merlin's father, Homer, was a logger, and his grandpa, Frank, ran sled yarder for Bill Meserve. His son, Jon, worked in the woods until he was injured. Merlin married Judy Kinney in 1964, and they have three children.

Ted Swanson

Theodore Willard "Teddy" Swanson

During World War II, Teddy was in high school. It was hard to find people to work because most men had jobs or were in the service. Teddy's dad, Ted Sr., needed someone to blow whistle for him. Teddy was eager to quit school and start his career in the logging industry, and his dad agreed. Teddy went to work and lasted only two days. He thought, "This isn't for me," so he quit and went back to school. Teddy had been working on the log booms rafting logs for several summers and had saved up his money to buy a car. He found a used one in Naselle. His dad took him to look at it but thought Teddy should buy a new car, so he did. He bought a 1946 Ford coupe that made him the big man at Naselle High School.

After he graduated in 1948, he went into the Air Force for four years. Teddy's dad was co-owner of Grays River Logging with Jim Oatfield. Teddy planned to go to work there, but instead Teddy and Darrel Clark bought O'Connors Inn and he worked there, but didn't stay long. Then he decided to try logging.

When Teddy's dad sold out his logging operations to Jim Oatfield, he gave Teddy, Gloria, and Lillian each one-sixth of Grays River Log. These three were involved in working for the company.

On a very stormy day, Teddy was at the landing, waiting for a load. The second loader was on the landing also. Teddy rolled his window down a few inches and told the second loader, "You got to be tough to be a logger." He wasn't too popular with the second loader for a while so he kept his truck door locked.

Teddy reported logging got better with time with all new equipment coming out. His favorite truck was a 1983 Kenworth used for a low boy. When Grays River Logging shut down operations in 1983, Teddy bought the truck.

In his truck-driving career, Teddy had two runaways, but he was able to get the trucks on the shoulder of the road in soft ground and stop them. Regarding changes in the industry, he commented that there are many rules and regulations nowadays that most of the old-time loggers couldn't put up with.

Teddy and Sally Sue Manula married in 1951 and have four children.

Dean Badger. Photo by Diana Jordan

Dean Arthur Badger

After graduating from Naselle High School in 1956, Dean went to work for Grays River Logging, working out of Klint Creek. He told them he was eighteen when he was only seventeen. That might have worked, but he got hurt on the job, so he spent the rest of the summer sweeping the highway. He then went to work for Haskin from North Bonneville. He worked for Weyerhaeuser for one year, then Crown Zellerbach for twenty-four years working construction and running cat. The first two years at Crown he was a powder monkey.

He decided to change occupations, so in 1980 he went to Alaska fishing. He now works at the Saturday Market in Ilwaco where he has a booth.

Dean was born in Longview in 1938. He has three children and two grandchildren.

Ed Wirkkala. Photo by Diana Jordan

Ed Wirkkala

Eddie went to work for Brix Logging one week before graduating high school in 1942. In 1943, Eddie went into the service and was in Europe for three years. When he came home, he worked for Wirkkala Brothers in Grays River on the cutting crew. In his career he was primarily cutting for the gyppo loggers on contract. In 1966, he went back to Weyerhaeuser to finish his logging career and stayed with Weyerhaeuser until they closed.

Eddie finished his working career at the Naselle School mowing lawns. Now he spends his days visiting family, friends, and working in his yard.

David Oatfield

David Oatfield

David grew up in Skamokawa, graduating from Wahkiakum High School in 1954. He was brought up on logging stories and was out in the woods helping his father, Jim, at an early age. Jim Oatfield was the owner and operator of Grays River Logging. David didn't think logging was for him, so he went to college where he met his wife of forty-five years, Donna Scotvald. They had three daughters. David taught school at Ilwaco for a few years, but felt the call of the woods and the desire to be following in his dad's footsteps. Back to logging he went, and loved every minute. He ran cat most of the time until he retired.

Everyone loved David. He had a great personality and loved poking fun at himself. In September of 2002, he passed away. As he put it about his father, Jim, when he passed away, "Dad is in 'Old Growth Land,'" and David is right there with him.

Clyde Sprague

Clyde Sprague

Clyde started logging in 1950 for Cap Schmand for $1.67 an hour. Clyde feels logging is still the same as it was, only different. Now there's little manual labor as it's mostly done by machine. Clyde logged for Jim Lindos and Grays River Logging. In 1960, he became a gyppo logger himself, logging for Olympic Hardwood, Weyerhaeuser, and Puget Sound Hardwood. He then logged for Snohomish Sawmill, the last big mill on the west coast. They could cut timbers sixty-five feet long. Clyde enjoyed logging wherever he worked. In his thirty-one years as a gyppo logger, he never had a man killed. That made every day and every year special to him. He employed a lot of "tramp loggers" because they understood logging and were excellent workers, even though he often had to bail them out of jail on Monday mornings.

He remembers that one of the big logging camps brought in a dietician to make sure the crew was getting plenty to eat and the right foods.

Clyde considered himself a "tramp clear-cut gyppo logger" because he logged in so many different places in Washington. His maternal grandfather was Martin Olsen from Finland. He was a timber faller in the late 1800s, working out of Deep River. His father, Chester Sprague, also worked in the woods as a rigging man and a head loader.

Clyde now lives in Boistford with his wife, Charmalee McKinnon. They have three children.

Joe Florek and Sons

Joe Florek

Joe Sr. started logging for himself right out of high school in 1944. He logged private timber, starting with his neighbors. When times got tough and he didn't have much to log, he would log timber on his home place. Joe stayed busy logging for friends and family until 1980 when his boys John and Joe Jr. took over. Kenny and Jerrold have also worked in the timber industry most of their lives. Jerrold was tragically killed in September of 2010, falling timber in Oregon. Jim is the only son who didn't become a logger.

Joe Sr. and Corliss Sotka were married in the early fifties and had eight children. Most of them graduated from college. Joe Sr. passed away in 2002. A memorial bench is located at the entrance to the county courthouse in Cathlamet, placed there by his family in honor of his years of leadership as a county commissioner.

Howard "HR" Nelson

Howard "HR" Nelson. Photo by Diana Jordan

Howard Nelson was born in Nebraska in 1931. "I remember working for my dad when I was six years old in Dillard, Oregon. I carried his oilcans for him when he was cutting old growth timber. I went to work as a log scaler when I was sixteen, then was in the Navy for four years. Drove truck in Oregon, then moved up here in 1965. I worked for Crown till 1985. Was a timber faller for many years, but my knees gave out, so I went to driving and I always liked it. Took a lot of years for me to get tired of it.

"Only time I ever got scared was when the brakes would go out. You had to hang on when you lost brakes and try to get the truck to drop off into a ditch to stop. My wife, Arlene, and I along with our six kids lived in Deep River part of the years when I worked for Crown. I drove off-road trucks and dropped logs at the Deep River dump. Those off-road trucks could really haul.

If Crown had been able to keep going, I believe we'd have plenty of timber today. They came in right after cutting and planted trees. They cut the ones that were ready to be cut, not every single one. 'Trees Forever'—that was the Crown motto. My happiest days were driving for Crown. I just liked driving trucks. As a matter of fact, I just drove for my son-in-law, Bob [Torppa], the other day. He didn't have anyone else that day, so I did it."

CHAPTER 9

DeBriae Logging Survives Industry Turmoil

by Jim LeMonds

Jerry was a great logger, one of the best around. He did it right.

Cliff Kilponen

Jerry DeBriae's office in Cathlamet tells you who he is—part logger, part CEO. He sits behind an expansive oak desk piled with papers and files, but you get the feeling it's just a prop. Branding hammers from previous jobs stand in the reception area. Caulk boots lean against the wall in one corner. The front third of a massive five-point bull elk appears to have crashed through the upper part of the wall near the door. Unopened bottles of Pendleton whiskey line one shelf of a floor-to-ceiling bookcase. He wears a hickory shirt, DeBriae suspenders. Gray hair peeks from under a camouflage Pape cap. When he moves, his body shows his seventy years, but his eyes are bright and it's immediately clear that he is still very much in the game. After more than a half-century of highs and lows, he is among the last independent logging contractors still standing in Wahkiakum County.

Jerry DeBriae, 2011. Photo by Bertroch

Jerry was born in 1940 at the Long-Bell company town of Ryderwood, Washington, the largest logging camp in the world. His father, Raymond, worked as a machinist in the Long-Bell shop. With nearly three thousand employees, spouses, and children, Ryderwood fed logs to the giant Long-Bell mill, fifteen miles south in Longview. When Ryderwood shut down in 1953, Raymond found work with Crown Willamette and the family relocated to Cathlamet. The DeBriaes have maintained a working relationship with Crown and its descendants ever since.

Jerry's father, Raymond DeBriae, at left, in Ryderwood in 1952. The woman on the right is Jerry's mother, Marie. Raymond worked for Long-Bell for nineteen years, then another nineteen years for Crown. Photo from DeBriae Collection

Jerry attended Wahkiakum High School and graduated in 1959. At the time, the timber industry was booming. "I wasn't much of a student," he said. "When I finished school, I needed a job, so I went to work in the woods." He's been there for more than fifty years. Those were the days of big trees, a booming domestic housing market, limited environmental restrictions, and plenty of work for loggers throughout the Pacific Northwest.

The Washington State Department of Resources had timber sales available in Wahkiakum County, and Crown Zellerbach was in the process of logging the last of its old growth at Brookfield and on the Elochoman River. Crown maintained a union crew into the 1980s, but there was still steady employment for gyppos like George Oller, Rueben Penttila, Kenny Martin, Ray Burton, Wayne Bond, and Jim Oatfield.

Jerry went to work for Crown in 1959, picking up rails off the old line that ran from Cathlamet north to Crown's Camp 2, which had closed the previous year. Constructed in 1929, Camp 2 included eighty-four buildings in its heyday and was the last active logging camp in Wahkiakum County. After that, he set chokers for Joe Florek before working on the rigging for local gyppos Durrah & McCormick from 1960 through 1966.

Jerry DeBriae on the right in the company's early years with a skidder and a yarder.

Penttila truck comes out of the mainline road in Deep River. Photo from Archives Collection

This Penttila truck now belongs to DeBriae Logging. Photo from DeBriae collection

In 1967, Jerry and brother-in-law Bob Olsen started their own operation—Olsen & DeBriae Logging. They first bought a Skagit SJ-4R yarder from a dealer. Two years later, they purchased a Skagit SJ-3R yarder from contractor Kenny Wirkkala and got a job decking logs for Crown.

After the partners went their separate ways in 1973, Jerry continued to run his own operation—Jerry DeBriae Logging. He had a cat and skidder and kept two sides going on a series of Crown contracts. He took a big step forward in September 1976. "I wanted to get into tower logging," he said, "and Reuben Penttila was selling out, so I bought one of his sides." The purchase included two Mack trucks, a Skagit BU-80 yarder, a Link Belt LS-98 loader, and TD-24 and TD-20 International cats. In addition to the equipment, Jerry continued to employ six men from Penttila's crew. The Skagit was equipped with inch-and-a-quarter mainline and three-quarter-inch haulback and had a reach of 1,000 to 1,100 feet. They were flying inch-and-an-eighth chokers. The timber was good-sized and included old growth hemlock, 30" to 36" in diameter. Jerry's crew trucked most of the logs to the Dant & Russell Mill at Warrenton, Oregon, or to the Crown dump at Deep River. He was thirty-six, and DeBriae Logging was going full-bore.

The expansion was not painless. Jerry had been a logger his entire adult life, but managing a business with dozens of employees and millions of dollars in overhead took some getting used to. "It took me a while to figure out how to run the business," he said. "That's part of the reason we didn't make much money until the 1990s." He also learned to delegate. "You have to have people you can trust," he said, "because there's no way you can do everything yourself." He'd run shovel during the early years of the operation, but by the early 1980s, he was spending most of each day in his pick-up. "There got to be so much stuff going on that it took all my time just to keep an eye on it," he said.

Jerry continued to expand. In summer 1980, he bought out Castle Rock gyppo George Oller, whose equipment included a tower, loader, and two trucks. "I kept his crew and finished the job he was working on for Crown out of Grays River," Jerry said. "That was a good one. By the time we were done, I'd made enough money to pay for all the equipment."

The landscape changed dramatically for nearly every logger and timber-dependent community in the Northwest during the 1980s, as a number of factors conspired to reduce employment in the wood products industry. In 1900, Weyerhaeuser Company had purchased nearly a million acres of Northwest timberland for $6 per acre. In Cowlitz County, Weyco worked east from present-day Interstate 5 toward the Cascades. By the late 1970s, the end of the old growth was in sight, though still a decade away. The May 18, 1980, eruption of Mount St. Helens changed that. Virtually all of Weyerhaeuser's remaining old growth on the St. Helens Tree Farm was now on the ground. It needed to be salvaged immediately before decay took hold.

Cy Robinson puts a new choker on Jerry's Skookum drift carriage. Photo from DeBriae Collection

Unidentified logger at a large cedar tree. Photo from DeBriae Collection

By the end of 1982, company crews—supplemented by a large number of contractors—had removed 850 million board feet of timber from the blast zone. The salvage operation had been a great success, but the paradigm had shifted. The old growth was essentially gone. And not just for Weyerhaeuser. Patches would remain into the 1990s, but suddenly the writing was on the wall. The next harvests would be much different. Plantation trees 12" to 24" in diameter would be the norm. The powerful equipment needed to move old growth would no longer be practical. In fact, with timber dimension so much smaller, machines could do much of the cutting.

Those sawmills that retooled to handle small logs had the best chance of survival; those that failed to upgrade were closed within a few years. Everything was focused on speed and precision, and the new equipment meant fewer employees would be needed.

In 1983, Weyerhaeuser announced that 1,100 loggers and millworkers would be laid off. The company also offered a buy-out to 250 loggers on the St. Helens Tree Farm in Cowlitz County. The package included a $3,000 severance

Aerial view of a DeBriae logging crew near Camp 2 on the 800. Photo from DeBriae Collection

payment and an annual pension—to be drawn starting at sixty-five—that totaled $300 per year of service. Discord continued until 1986, when 7,500 Northwest loggers and millworkers who were members of the International Woodworkers of America walked out in the face of a company proposal to cut wages by 25 percent. The company held fast, and the union relented.

Weyerhaeuser's move to smaller trees, faster and lighter equipment, and fewer employees was mirrored by timber companies throughout the Northwest. Weyco employed nearly two hundred union loggers in Wahkiakum County until 1983 when the company's Grays River operation shut down. A company spokesman said Weyerhaeuser's remaining timber was mostly hemlock, which lacked market demand and profitability.

"They called the union boys together one day," recalled lifelong Wahkiakum County resident Kent Martin. "They told them the good timber had been logged out and their jobs were over. I remember going to a school event for my kids in the afternoon and seeing more men there than I'd ever seen before. It was the logging crews who had been laid off. They had no work and showed up to watch their kids." Crown took the same tack as Weyerhaeuser.

Jerry and his wife, Linda, with a new truck circa 1980s. Photo from the DeBriae Collection

In *Beach of Heaven*, Wahkiakum County author Irene Martin notes that Crown employed approximately 260 wage and salaried workers from 1961 to 1981. "Force reductions began in 1981," Martin said, "and by 1985 only 97 employees remained." When Sir James Goldsmith purchased Crown in 1985, the company ended its union operations.

Jerry was one of the few Pacific Northwest loggers not affected by the eruption of Mount St. Helens. While many other gyppos signed on to help Weyerhaeuser log blow-down in the blast zone, he continued to work for Crown, primarily in Wahkiakum County. "The only thing that hurt us," he said, "was that Weyerhaeuser put so much wood on the market it screwed up the prices for a couple of years."

After Crown and Weyerhaeuser axed their union crews, unemployment in Wahkiakum County hit 34.6 percent. Some work remained available for independent contractors, who filled the void when union crews were cut. However, Crown operations continued to endure changes that made life uncertain. When Sir James Goldsmith took control of the company, he immediately began parting it out. James River got the pulp mills while the timberlands went to Cavenham. In the early 1990s, Campbell Group (a Timber Investment Management Organization) took over management of the former Crown land.

Jerry continued to look for new opportunities. By 1988, he was running four tower sides and several shovel sides. The expansion continued in 1990 when he partnered with Sonny Lundberg to form S & J Logging. "We also got into road construction about that time," Jerry said, "and that took off. We were probably building about twenty miles of road a year." Hampton had purchased a number of state sales in the early 1990s and sent plenty of business to S & J Logging. "We were doing $80 million a year just off Hampton," Jerry said. "We also did a lot of work for Campbell Group and Boise-Cascade at that time." Jerry logged on the Kalama River for Boise-Cascade while Sonny ran another operation for Boise in Oregon. They continued to grow, and in 1997, '98, and '99 did more than $1 million per year in gross sales.

Crown had operated a sorting and chipping operation in Cathlamet for many years. When Cavenham Forest Industries purchased Crown's timberlands, Jerry and Mark Linquist bought the sorting and chipping business, which they ran from 1989 to 2000. At its peak, the yard was handling fifty to seventy-five loads of logs and chips per day. "We bought a chipper that was a key to keeping the sorting yard running," Jerry said. "That thing was a gold mine." Export logs were rafted to Longview or Astoria. The chips were barged across the Columbia to the James River paper mill on the Oregon side at Wauna. Jerry and Linquist signed a contract that was supposed to run through 2016. "But," Jerry said, "Campbell Group [which had purchased the yard from Crown] found a loophole and started trucking the wood directly to the mills in Longview."

When Sonny Lundberg died in 2000, Jerry purchased Sonny's share of S & J Logging. By 2001, Jerry was employing 125 people and running seven towers, twelve de-limbers, and twenty-three shovels. The company maintained a fleet of thirty trucks and utilized the services of forty gyppo rigs. The 2001 issue of *Timber/West* magazine featured the top ten logging contractors by volume in Washington, Oregon, Alaska, Montana, Idaho, and Northern California. DeBriae Logging led the way with an incredible 125 million board feet per year. That year marked the company's peak production.

Jerry points to improved technology and smaller timber size as the biggest changes in the timber industry during the past half-century. "You aren't going to see a tree over forty years old these days," he said. "The timber companies' investors want their money, and they don't want to have to wait for it." Jerry's crews had their last shot at logging old growth during the early 1990s while working for Boise-Cascade on the Toutle and Little Kalama Rivers.

Now the trees are much smaller, and that has necessitated a change in equipment. "Towers are still towers," he said. "It's just that today you're logging much smaller timber. It's the switch to mechanical logging that's the big change." Yarders are engineered much differently today. "It's all about speed," Jerry said. "In the old days, the logs you were pulling were much bigger, and there were big stumps in the way."

A delimber at work in upper Salmon Creek on a DeBriae cutting site in 2011. Photo by Bertroch

In the days of big timber, yarders had a reach of approximately one thousand feet. Today, with trees averaging just 12" to 24" in diameter, each turn (group of logs yarded to the landing in a single pull) is so light the reach can be as far as two thousand feet. There's no need to worry about a big pull in a low gear. The key is to get turns to the landing as quickly as possible.

Jerry ran a grapple shovel during the 1960s, '70s, and '80s. He can't imagine doing the loading today without hydraulics. It takes many of today's logs to generate a load, so speed and profitability go hand in hand.

He began the switch to feller-bunchers, processors, and hydraulic shovels during the 1990s. Mechanical logging—which does not rely on humans to cut the trees—now accounts for approximately 80 percent of the company's output. Feller-bunchers, which snip trees at the base and lay them on the ground, can operate on 40 percent slopes. Jerry has utilized as many as twenty hand-fallers in the past, but that number was down to four in 2011. They handle the cutting where the terrain is too steep for machines. Jerry estimates his crews produce five more loads per day on sides that can be mechanically logged, as compared with those that are logged using the conventional tower method. But the price is steep. He owns three feller-bunchers; they cost $400,000 apiece.

Processors, which handle de-limbing and bucking, are equipped with sophisticated computers that assist in bucking the logs to the proper length. Jerry owns nine, at a cost of $500,000 per machine. "It seems like they wear out just about the time you get them paid for," he said. More mechanical cutting means fewer employees

Jerry always takes his dog to work with him. His dog, Chip, occupies the entire back seat area of the truck, and his dog food is in a plastic bag beside Jerry. Chip is well trained as a retriever, as all Jerry's dogs have been. He prefers black labs and has had several over the years. Here, Chip is barking to get Jerry to throw a stick for him—which he did. Photo by Bertroch

on the ground. And that translates into lower costs for state industrial insurance. In fact, Jerry's rate was one of the lowest in the state among independent logging contractors for several years.

In 2010, he paid $15 per hour in state industrial insurance for each timber faller he employed. Meanwhile, the cost was just $1.06 per hour for employees using feller-bunchers to cut timber. DeBriae Logging bought three new trucks per year from 1997 through 2007, but tougher economic times have put new purchases on hold. "A new truck runs about $150,000," Jerry said. "You just can't justify buying one in today's economy." Because the truck fleet is aging, DeBriae Logging puts a big emphasis on maintenance. "We handle the engine and transmission work in our own shop and send out the other stuff," he said.

Jerry looks over the upper Salmon Creek area. Photo by Bertroch

Jerry checks out a work site in the upper Salmon Creek area, April 2011. Photo by Bertroch

Loggers conformed to very few environmental regulations during the 1960s and '70s. Trees were cut right down to the stream bank, equipment was run through fish-bearing creeks, and road construction was inadequate to deal with run-off. Things are much different today. "It's just a lot more complicated," Jerry said.

During the twentieth century, heavy rainfall led to shutdowns only when saturation made it impossible for equipment to operate. Today, rainstorms can generate a call from the State of Washington, ordering companies to halt logging and road-building operations until things dry out. The problem? Sediment that makes its way into streams has been found to degrade water quality and cause extensive damage to fish habitat. Studies conducted during the second half of the twentieth century clearly proved that poor logging and road construction practices increase run-off and damage fish populations.

A 1998 report by Larkin, Slaney, Warburton and Wilson titled "Suspended Sediment and Fish Habitat Sedimentation in Central Interior Watersheds of British Columbia" echoed what

Logs at the upper Salmon Creek work site, April 2011. Photo by Bertroch

other researchers had been saying for several decades. The report found that "Excessive stream sedimentation is a prevalent problem in watersheds damaged by poor forest land use practices, and affects both spawning and rearing habitat of salmonids, and their aquatic food chain."

For example, the annual water volume in Carnation Creek (British Columbia) increased 9 to 16 percent after the watershed was clear-cut. A study conducted in western Oregon found that clear-cutting increased water discharge rates by as much as 50 percent in small basins and 100 percent in large basins. "The State has really

Marvin Paavola cuts a large spruce tree in the Chinook watershed. Unidentified man on the left is assisting with hydraulic jacks and wedges. Photo from the DeBriae Collection

tightened things up when it comes to run-off," Jerry said, "especially if you're working around fish-bearing creeks."

Logging can result in water quality problems, but road construction has the most potential for damage. Today, building road means knowing the rules. For that reason, Jerry and one other experienced employee handle what construction work there is. DeBriae Logging still does four to five miles of road-building per year, but with so little logging going on, few timber companies are willing to spend money on road construction projects.

Washington's Forest & Fish Law, passed in 2001, is focused on improving road maintenance and construction practices to reduce run-off into streams. Culverts and bridges are larger in size than ever before and new road-building methods are now in place. Some roads are built at a perpendicular angle, instead of parallel to streams. This reduces the amount of road surface that can contribute sediment. New cross-drain techniques divert run-off from ditches onto the forest floor, and sediment traps are put in place to prevent sediment from reaching streams.

When a road is graded or sloped, new strategies are being used. The surface might be crowned, with a high spot in the middle of the road, allowing the water to run off to either side. Where practical, a road might be sloped outward to tilt the surface toward the downhill side for improved drainage.

Most people think the spotted owl is the bird that has had the biggest impact on logging in the Pacific Northwest, but in Wahkiakum County the culprit is the marbled murrelet. Dependent on old-growth forest for nesting, the marbled murrelet suffered a significant fall-off in numbers during the twentieth century. The U.S. Fish & Wildlife Service declared the murrelet a threatened species in Washington, Oregon, and California in 1992. "Not much logging has taken place on DNR [Washington Department of Natural Resources] land at the Coast or around Naselle for years because of the marbled murrelet," Jerry said. Marbled murrelet restrictions have virtually no effect on logging on private timberlands, however, because the trees are far too small to be considered old growth.

When timber prices fall, Pacific Northwest schools and county governments feel the sting. In 2009, timber prices were down nearly 60 percent from the 1990s. When Washington was granted statehood in 1889, the federal government designated two sections of land in each township as revenue-generators for public schools and universities. Some states sold off their holdings, but Washington did not and now has more trust land than any state in the nation. Funding from trust land has also been used to support rural governments.

The Washington Department of Natural Resources (DNR) manages 5.6 million acres, of which 2.1 million acres are timberland. DNR's holdings account for 12 percent of the commercial forestland in Washington, and the agency generates approximately 17 percent of the annual harvest. Jerry DeBriae Logging counts on state sales for nearly 60 percent of its revenue.

One sticking point is that the state's constitution stipulates, "public lands granted to the state are held in trust for all the people." Some environmental groups and recreation advocates say that state lands are generating funds only for public schools and universities and county governments. Their increasing pressure has forced DNR to rethink its mission and to back away from the notion that state timberlands are solely dedicated to schools and county governments.

The situation is dire in Wahkiakum County. With little industry or federal land to provide revenue, 40 percent of Wahkiakum County's budget is generated by DNR timber sales on state lands. Without this money, the county has difficulty maintaining jails, roads, and a police force. Currently, market prices for timber are much lower than they were in the 1990s. This leaves DNR in a quandary. Does it hang on to

Jerry's lunchbox in his truck. It's seen many years of service. Photo by Bertroch

its timber until prices climb, or does it harvest at reduced value to generate funding for schools and county governments? The Washington Department of Natural Resources generated $230 million in timber revenue in 2006. The agency expected that number to be closer to $107 million in 2010 and 2011.

Oregon is in a similar situation and expects to see a fall-off of nearly 35 percent in timber revenue for the state's common school fund. Things are so bad that timber companies are reluctant to bid on state sales—there's simply no profit to be had.

Jerry said the past couple of years have been as bad as he's ever seen. In fact, he shut down operations for four months in 2009, which put nearly one hundred employees out of work. In an interview with Northwest Public Radio, he said, "You couldn't sell a log. You couldn't sell anything. There was no demand. There's [*sic*] so many people out of work they bid the jobs down to nothing." There is general consensus that it is no longer practical to expect state timber sales to fund schools and county governments. The question is this—Where do you get the dollars to provide adequate, reliable funding? There is no answer in sight. Times have been tough in recent years, not only for Jerry but for loggers throughout the Northwest.

The U.S. housing market peaked in 2005 and began to decline in 2006. As of 2011, it had not hit bottom. With the number of housing starts in decline, there has been virtually no domestic demand for lumber. The impact has been widespread. Jerry DeBriae Logging failed to make money for the first time in 2009 and 2010. His crew has been reduced 30 to 40 percent from the late 1990s. "We're working for less money now than we were five years ago," he said. "We provide benefits to our employees, but we'll probably have to cut back what we offer. We're fortunate to have the Chinese market because it's the only thing keeping us going right now."

In response to a global recession, China took action to stimulate its economy by increasing log imports from the United States to meet demands for housing and manufacturing construction. With its own resources unable to keep up with demand, China has relied increasingly on Oregon and Washington to provide Douglas fir and hemlock.

According to Bob Flynn, director of RISI, market focus has shifted among log exporters in the Pacific Northwest. "In the first quarter of 2008, only 2 percent of the log exports from this region went to China, compared with 40 percent to Korea and 59 percent to Japan. But in the first quarter of 2010, the share of log exports going to China increased to 23 percent of total log export volume, compared with 32 percent to Korea and 45 percent to Japan."

The *Daily News* (Longview) reported that log exports at the Port of Longview and Weyerhaeuser docks were up 60 percent from 2009, while wood chip exports rose 25 percent. The increase was tied directly to increased demand in China. With the U.S. housing market on life support, it is likely Jerry and other logging contractors will be dependent on China for some time. "In some years we made a lot of money," Jerry said, "but the best thing about being in this business is that we've kept a lot of people working in Wahkiakum County."

Jerry checking in with one of his operators.

He has four sides going now and hopes to secure enough work in 2011 to expand to six or seven. He still has twenty trucks of his own, but it is increasingly difficult to keep them all working. With state timber sales virtually off-limits, he continues to work for Hancock Timber Resource Group, which now manages the former Crown Zellerbach timberlands for a group of investors. "They've been good to deal with," Jerry said. "Unlike some of the others, they'll let you make a dollar if there's a dollar to be made."

He's done the vast majority of his logging in Wahkiakum County, but today's economy means keeping an open mind about working elsewhere. "I like to keep my jobs within one hundred miles of home," he said, "but when you have all this equipment to pay for that isn't always possible." Not only is the equipment different these days, so are the people. "In the old days, loggers were loggers," he said. "Nowadays, it's tough to get good workers. Heck, it's hard just to find people who want to work in the woods."

Asked how much longer he planned to stay in the game, Jerry said, "That's a good question. If the market picks up, maybe we'll keep at it." Jerry's son, Jerry Jr., has worked with his dad since the mid-seventies and now handles equipment maintenance and the trucking side of the operation. He is also in charge of moving towers and other equipment from one job site to another. He will eventually run the company. "He does most of the woods work, and I do the paperwork," Jerry Sr. said. "He'll keep it [the company] going."

Jerry vacations three to four months per year and has hunted and fished in Africa, Argentina, New Zealand, and Australia. In January 2011, he had just returned from a fishing trip to Costa Rica and plans another trip to Africa in 2012. When he's out of the office, he carries a cell phone and stays on top of company business. He might be winding down, but he's not ready to give up the job he has known and loved for so many years.

"I guess I need some place to go in the mornings and something to do," he said. "I probably won't get out until they put me in the ground."

A yarder is being moved by two cats, one pulling and one pushing. Photo from the DeBriae Collection

EPILOGUE

A Final Word

by Nineteenth District Washington State Representative—and logger—Brian E. Blake

The timber industry has changed. We've moved from pioneers clearing space for farms and producing lumber for growth, to tax laws landowners can't afford. Historically, the integrated timber companies fostered community stability. We've moved away from company towns to a new world with international investors. We're all learning to cope with the changes.

Logging was and still is an honorable job. Like so many others in this last frontier state, I started out as a logger. I had my beginning in the woods as a choker setter. Things go smoothly unless you get careless; then accidents happen. I've been involved in a couple of accidents, one of which resulted in two of my friends dying. While my carelessness was not the cause of their deaths, accidents like that make one more careful than ever—sometimes there are no second chances. Logging today is as dangerous as ever.

In my experience, loggers consistently go home wishing they could get one more load of logs into the mill. We take great pride in our work and naturally want to do better each day.

Upon starting logging, I became aware that loggers are diverse; through their examples, I realized the work must be taken seriously every day. Production was a natural goal for loggers because we took pride in giving a day's work for a day's wage. This is just such a reflection of the settlers of Washington State. They settled here because they valued the natural resources so plentiful in the streams, valleys, and forests. Loggers today value our resources no less than the early loggers. I believe we may honor our resources so deeply because we know what can happen when we fail to pay attention.

Wahkiakum County Commissioner and timber faller Dan Cothren says that logging is a different job now, but the same spirit. I agree. Logging is not only an honorable profession; it is an honorable life for the logger's family and community.

In the communities where we lived, we understood the importance of taking care of trees. We knew each logging road as well as our own community streets because we had access. Now with the majority of timberland belonging to private investors rather than local or regional landowners, there is little access allowed. This is one reason I originally ran for office, my concern regarding public access to our forestlands.

We still have amazing diversity in our forests. The new stream buffers are working. I'm pleased with how well our state agencies are developing forest and fish practice policies that work for the environment,

the people, and for the timber industry. I'm concerned, however, that we have a lack of connection between conservation groups, government, and our timber communities.

It is my hope that in the coming years, we will see greater understanding from our urban activists about the need for continued sustainable management and stable rural economies.

State Representative Brian Blake, right, with Wahkiakum County Commissioners, from left, George Trott, Mark Lindquist, and Dan Cothren, on a timber tour in 2003. Three of the four electeds are, or at one time were, loggers. Photo by Bertroch

INDEX

D

E

F

G

H

I

J

K

L

M

N

O

P

Q

R

S

U

V

W

Y

Z

About the Authors

Karen Bertroch, director of the Appelo Archives Center, has been a grant writer for more than thirty years. She has had articles published in magazines and developed training curriculum for nonprofits for her business, Faith Leadership Consulting. She loves history and is intrigued by local history and the Finnish influence on southwest Washington communities. Research for this book was a labor of love for her.

Donna Gatens-Klint, a native of the southwest Washington area, was born and raised in Nemah and attended school in Naselle. In 1958 she married George Klint, a farm boy from Grays River. They have two daughters, Janine Klint Davis, who teaches school in Cathlamet, Washington, and Diane Klint Brown, a registered nurse. This is Donna's third book.

Jim LeMonds is the author of *South of Seattle* and *Deadfall: Generations of Logging in the Pacific Northwest*. A former teacher, he now runs his own marketing business. He lives in Castle Rock, Washington. Additional information about Jim is available at www.writeteknorthwest.com.

Bryan Penttila, a logger by profession, has been interested in the workings of the region's forests for most of his life. He graduated from Eastern Washington University with a degree in history and spent several years in the graduate studies program at Washington State University. He and his wife, Jennifer, live in Naselle, Washington.

Bill